TAKING THE PLACE OF FOOD

TAKING THE PLACE OF FOOD

Khat in Ethiopia

Edited by
Ezekiel Gebissa

The Red Sea Press, Inc.
Publishers & Distributors of Third World Books

P. O. Box 1892　　　　P. O. Box 48
Trenton, NJ 08607　　Asmara, ERITREA

The Red Sea Press, Inc.
Publishers & Distributors of Third World Books

| P. O. Box 1892 |  | P. O. Box 48 |
| Trenton, NJ 08607 | | Asmara, ERITREA |

Copyright © 2010 Ezekiel Gebissa
First Printing 2010

All rights reserved. No part of this publication may be reproduced, stored in a retrieval system or transmitted in any form or by any means electronic, mechanical, photocopying, recording or otherwise without the prior written permission of the publisher.

Book design: Saverance Publishing Services
Cover design: Ashraful Haque

Library of Congress Cataloging-in-Publication Data

Taking the place of food : Khat in Ethiopia / edited by Ezekiel Gebissa.
  p. cm.
  Includes bibliographical references and index.
  ISBN 1-56902-317-4 (hardcover) -- ISBN 1-56902-318-2 (pbk.) 1. Khat--Social aspects--Ethiopia. 2. Khat--Economic aspects--Ethiopia. 3. Khat--Political aspects--Ethiopia. 4. Cash crops--Ethiopia. 5. Ethiopia--Economic policy. I. Gebissa, Ezekiel.
  HV5822.Q3T35 2010
  362.29'9--dc22
                          2010003444

Table of Contents

෨෬

List of Tables and Figures	vii
Preface	ix
Chapter 1 | Introduction: Ethiopia's Khat Dilemma *Ezekiel Gebissa*	1

PART I: CULTURE OF CONSUMPTION

Chapter 2 | Tradition and Innovation in the Ritual of Khat Consumption in Wallo, Northern Ethiopia *Hussein Ahmed*	13
Chapter 3 | Chewing and Dreaming: Youth, Imagination, and the Consumption of Khat in Jimma, Southwestern Ethiopia *Daniel Mains*	29
Chapter 4 | Keeping Tradition and Killing Time: The Use and Misuse of Khat in Ethiopia *Ezekiel Gebissa*	57

PART II: ECONOMICS OF PRODUCTION AND TRADE

Chapter 5 | Crop and Commodity: Economic Aspects of Khat Production and Trade *Ezekiel Gebissa*	89

Chapter 6 | Agrarian Debacle and the Spread of the
Dollar Leaf in Northern and Southern Ethiopia   127
Degol Hailu

Chapter 7 | Khat and Livelihood Dynamics in the Harer
Highlands of Ethiopia: Significance and Challenges   149
Habtemariam Kassa

Chapter 8 | Market Incentives, Rural Livelihoods, and
a Policy Dilemma: Expansion of Khat Production in
Eastern Ethiopia Beyond the Tesfaye   167
Tesfaye Lemma Tefera and Daniel Start

PART III: POLITICS OF POLICYMAKING

Chapter 9 | Beyond the Politics of Prohibition
Ezekiel Gebissa   191

Chapter 10 | Afterword   201
Christopher Clapham

Bibliography   209
Notes on Contributors   229
Index   231

List of Tables and Figures

TABLES

5.1:	Production: Land use trend in Harerge, 1954-2000	103
5.2:	Khat cultivated land and production levels, Eastern Ethiopia, 2000.	105
5.3:	Cropping patterns in Haramaya, 1991-2003	105
5.4:	Changes in the retail price of khat per kg, 1970s–2004	110
5.5:	Comparison of income for khat growers and non-khat growers in birr	115
5.6:	Food security indicators for khat and non-khat growers	116
5.7:	Changes in khat income tax, 1965-2004	123
6.1:	Khat production and utilization in Ethiopia, 2001/2002	133
6.2:	Area and production covered under extension program	135
6.3:	Taxes from khat and total tax revenue for 2001/2002, SNNPR	137
6.4:	Khat Tax: Volume and Revenue, SNNPR	137
6.5:	Area and production of permanent crops, Amhara Region, 2001/2002	140

6.6:	Amhara Region: Crop production and utilization, 2001/2002	140
6.7:	Bahir Dar: Tax revenue from khat in birr	145
8.1:	Cropped area shares for main food crops (percentages of AEZ/wealth categories)	173
8.2:	Crop choices and trade-offs matrix	175
8.3:	Gross income per unit area [=value yield] for main crops (birr per *quindi*)	179
8.4:	Livelihood poverty option map	181

FIGURES

4.1:	Administrative Regions	63
5.1:	Terraced mountainside khat plantations, Upper Dhengego, c. 1970.	100
5.2:	Terraced mountainside khat plantations, Lower Dhengego, c. 1970.	101
5.3:	Khat cultivation areas and regional states, 1990s	102
6.1:	The Southern Nations, Nationalities and Peoples Regional State	132
6.2:	The Amhara Regional State	138
7.1:	The crop-livestock system model indicating the role of khat in the system	154
7.2:	Adaptive expectations of farmers while making production systems choices and management strategies	155

Preface

෫෬

It took a much longer period of gestation than I originally imagined for *Taking the Place of Food* to be published. Many of the chapters in this volume were initially presented at a workshop, "Khat and the Ethiopian Reality," held at Addis Ababa University in 2004. At the time, the presenters underscored the fact that khat had come to play an important role in Ethiopia's economy and society. The purpose was to draw policymakers' attention to the need for an effective khat policy. In retrospect, the intervening years since the workshop have only enhanced the validity of the argument underlying this volume—Ethiopia needs a policy for regulating the production, consumption and trade of khat. The case we tried to make at the workshop has now become even more pressing in light of growing international pressure to ban the trade and use of khat.

The chapters in this volume make it clear that khat is now an integral part of agriculture and supports the livelihood of millions of farmers in many parts of Ethiopia. Its trade provides employment for thousands of men and women who would otherwise swell the ranks of the urban unemployed. The revenue from taxes and duties on khat transactions supports government expenditures and development efforts and the cash flow generates multiplier effects in the regions where the cash crop is grown. In the traditional context, khat is used in ritual and habitual ways and plays a significant role as a social lubricant, cultural glue, and an identity marker. As such, khat is not a drug that leaves a trail of destruction in its wake in the form of addicts, criminal traffickers, and corrupt officials. The khat farmer, trader, and majority of chewers in Ethiopia are deeply tied to the leaf for historical, economic and cultural reasons. Its prohibition would likely cause tremendous economic and cultural dislocation.

Taking the Place of Food explains that the shift to high-value cash crop agriculture in many parts of Ethiopia represents farmers' informed and forward-looking response to changing market situations, political adversities, environmental challenges, and family considerations. The shift was inevitable in a situation where no alternative crop possessed khat's economic benefits. While outlining khat's economic importance and cultural significance, the chapters in this volume also stress that khat production cannot be the foundation of sustainable rural development in Ethiopia based on the efforts of smallholder producers alone. The authors highlight the stark choice that the farmers face between extreme poverty and the allure of a high-value cash crop. The country faces a similar dilemma associated with the sensitive issue of collecting tax revenues from transactions involving a substance many consider a harmful drug. It is a quandary that cannot be resolved with existing know-how or current expertise. Effective policy decisions require an understanding of the complexities and sophistication of highly diversified, farm-level agricultural knowledge and practices.

The chapters in this volume collectively argue that official policy thinking must move away from recycling various iterations of rural development strategies that perpetuate the doctrinaire assumption that smallholders must be assisted by the government to become self-sufficient farmers to emphasizing the need for diversification of the regional economy in which more people are engaged in nonfarm economic activities. At some point, policymakers and development experts must come to the realization that the country simply does not have an infinite supply of physical space able to accommodate more and more farmers.

Preface

The workshop where the idea of this volume was conceived was supported by a financial assistance from the Public Affairs Section of the United States embassy in Addis Ababa. I am indebted to the officials of the section at the time for facilitating the grant that made the workshop possible. By sponsoring the workshop, the Institute of Ethiopian Studies and the Office of the Vice President for Research and Graduate Studies at Addis Ababa University made possible the use of university facilities. I am particularly grateful to Prof. Endeshew Bekele, then vice president for research and graduate studies, for covering the expense of printing the proceedings of the workshop papers.

After the workshop was completed, I had to travel back and forth to Ethiopia to not only conduct research for my own contributions to the volume, but also to make sure the original contributors remained onboard and new contributors were identified to take the places of those whose track I had lost. This would not have been possible without the generous support I received from Kettering University. I want to express my gratitude to the university and my admiration to Karen Wilkinson, head of my department, for cutting through bureaucratic red tape and making the funds available every time I lodged my request for travel assistance. I am grateful to Bonnie Holcomb and Abebe Adugna for their interest in my work and comments on my chapters.

Finally, I would like to express a heartfelt debt to my family for allowing me to pursue my career at their expense. I hope that the final product compensates for the inconvenience that my frequent absences have caused them. I wish I could promise them that this is the last leg of my scholarly journey.

Ezekiel Gebissa
Kettering University, July 2009

Chapter 1

Introduction: Ethiopia's Khat Dilemma

Ezekiel Gebissa

෩෬

[Khat] is held by the Olema here [Harer] as in Arabia, 'Akl al Salkin', or the food of pious and literati remark that it has the singular properties of enlivening the imagination, clearing the ideas, cheering the heart, diminishing sleep, and *taking the place of food*.

Sir Richard Burton

Khat (*catha edulis*) is a psychoactive shrub grown in the Horn of Africa and the Arabian Peninsula whose leaves are consumed for their euphonizing effects. In Ethiopia, khat chewing was traditionally identified with Muslims, particularly those living in the eastern part of the country. Indeed Islamic religious leaders chewed the leaves during Ramadan (the Muslim fasting month) to stay awake for the long nights of prayer, merchants during long distance travels, and farmers for energy (Jaenen, 1956). Christians generally disparaged and consciously avoided the practice. Muslim women rarely chewed the leaf either. In the late nineteenth century, only the religious and political elite along with well-off urbanites of the city of Harer, considered one of the holy cities of Islam, were known to chew khat regularly (Burton, 1987, pp. 53-55).

The chew culture was long confined to the natural habitats of khat because the leaves lose their potency within two days of being cut and become unfit for use. Over the last half century, however, the introduction of improved motor and air transport has decreased the time of delivery of the leaves to consumers and consequently enlarged the area of consumption. Within Ethiopia, khat chewing has become a ubiquitous habit, cutting across class, religious, ethnic, and gender affiliations. A stimulant grown in small gardens for consumption on cultural and religious occasions at the turn of the last century, khat has now become the preferred cash crop in Harer and beyond, a highly sought-after substance to support a visible and pervasive social habit, and an important income-generating occupation for millions of Ethiopians. Ethiopian producers supply fresh leaves to chewers in Djibouti, the Republic of Somaliland, and several Middle Eastern countries.

During the 1980s, refugees from the Horn of Africa further spread the custom of khat-chewing to their host nations, including the United States, Canada, Australia, and various Western European countries. According to rough estimates, five to ten million people worldwide currently use khat on a daily basis. Though generally considered a psychoactive substance, khat has not been classified internationally as an illegal substance; it is legal in some countries, and banned in others. The Ethiopian government's attitude toward khat can best be described as ambivalent, on the one hand, content with the revenue khat generates, but, on the other, antagonistic to the consumption it sees as a purveyor of indolent behavior.

Khat consumption has become a phenomenon in both rural and urban areas in the Horn of Africa, and in large metropolises in Western Europe and North America. Production has increased to meet the expanding domestic, regional, and international demand. The amount of arable land devoted to khat cultivation has increased dramatically, as has its role in Ethiopia's international trade. In 1999, khat became Ethiopia's second largest foreign exchange earner, next to coffee, and has since remained one of the principal export crops of the country.

Many experts assert that the dramatic increase in production resulting in the vast expansion of khat fields has led to the displacement of food crops. Consequently, development workers and policymakers have expressed concern about the Ethiopian farmer's food security and self-sufficiency. The evidence from smallholder farmers, particularly from eastern Ethiopia, does not support the view that every inch of the available arable land should be devoted to food crop production in order to achieve food security or self-sufficiency. In fact, the smallholder farmers view the logic of agricultural extensification as the wrong path to food self-sufficiency. Over the last six decades, population pressure has reduced the amount of arable land in the region, accompanied by a severely diminished capacity on the part of the farmers to feed their households. In response, farmers began to show preference for mixed cropping as the more logical farming practice over the age-old grain-dominated cropping regime.

Even though non-farmers regard khat as a non-agricultural crop, many farmers grow the shrub extensively, largely for meeting their constantly increasing cash needs. Income from khat is considerably higher than from any other crop and is attained with minimum agronomic input. Unlike coffee, its rival cash crop, khat allows intercropping, thrives on marginal lands of low-level soil fertility, requires minimum technical and labor input, and is appreciably disease-tolerant. High income from cash cropping has proven to be the farmers' strategy for achieving food security. Farmers report that they currently operate on the principle that having *cash makes them more food secure than growing food crops* because it allows them to purchase food year round and additionally provides secure cash flow that can be used for investments in non-farm sectors.

Consequently, the cereal-coffee-livestock complex that has characterized Ethiopian agriculture for many decades appears to have been challenged by the high-income cash crop, khat. The shift to high-value cash cropping that began in eastern Ethiopia is being replicated in other parts of the country. In some cases, khat is planted on land that formerly had been reserved for food crop cultivation and pastureland. In other cases such as in the Jijiga Zone of the Somali Region in Ethiopia, the new farming system is an integrated agropastoral production system with khat cultivation providing most of the cash income for farmers. Household surveys indicate the prevalence of a much stronger view about the importance of mixed cropping for the farmers in some coffee/khat-growing midlands of south central Ethiopia such as the Wendo Genet region. Studies from eastern, southern, and northern Ethiopia now show that farmers believe that attempts to pursue household self-sufficiency by growing grains, as advocated by the government and some NGOs, actually reduced rather than improved food security. In fact, khat farmers, certainly responsive to market incentives, maintain that their chosen cash crop allows them to maximize profits and avoid risk and drudgery. Food constitutes the main expenditure of farming families who grow khat and other cash crops. One farmer interviewed on the matter stated that without khat the population in eastern Ethiopia would perish as a whole – both rural farmers and urban traders alike (Gebissa, 2004, p. 149). In view of khat's advantages, it is virtually impossible to convince smallholder farmers to halt the expansion of khat production using arguments relating to food security.

The choices farmers make are structured largely by the marketing of khat leaves. Apart from the employment opportunities it generates for a significant proportion of the society, khat marketing has economic importance for many agricultural enterprises, generates unprecedented tax revenues for regional governments, and provides a net return of much-needed foreign exchange for the country. The money multiplier effects of the cash accruing from the khat industry, as the chapters in this volume will show, is huge. Inevitably, the farm level decisions are heavily influenced by the larger political economy of khat.

The expansion of khat consumption, production, and trade has given rise to heated public debates. Over the last decade, print and mass media outlets have carried programs devoted to examining the "balance sheet" of khat use. Opponents of khat regularly campaign in favor of proscription, arguing that the practice of chewing promotes laziness, violence, and cultural decadence among users. Even educated people, who would otherwise base their opinions on empirical data, occasionally criticize routine khat consumption for its presumed adverse effects on national productivity and the health of individual citizens. Because of khat's traditional association with Muslims, some Christians have been heard bemoaning the loss of young Christians to an "Islamic culture." A few complain about the lack of proper context for a national debate since the positive economic and cultural significance of khat is often neglected.

Indeed the public discourse on khat is taking place in a context where casual observation passes for empirical evidence. Scholarly studies on khat have until recently focused on the medical and pharmacological effects of habituation, accentuating the alleged deleterious consequences of chewing on the consumers' physical and socioeconomic health. Researchers seldom attempt to understand the khat phenomenon from the perspectives of producers, marketers, or consumers. The national debate has thus been shaped by personal beliefs, cultural values, and narrowly constructed "scientific" studies, and has dwelt on the supposed adverse consequences of khat consumption apart from the economic significance, agronomic roles, and cultural significance. The fact that khat has become a critical commodity for the survival of millions of producers and marketers gets short shrift in public debates on the dramatic expansion of khat production and consumption. Consequently, the people whose livelihoods are connected with khat and who are most likely to be affected by the outcomes of the controversy have been excluded from the debate.

The palpable absence of adequate knowledge about khat in the debates underscores the need for a multidisciplinary study of khat production and the national and international trade, together with the impact of consumption on the user. Such a study would provide a clear understanding of the intermeshing of Ethiopia's political economy with the khat industry. By focusing on the relationship between khat production and markets, the causes and consequences of the cyclical reinforcement of a regime of controls on agricultural marketing, we will gain a better understanding of the socio-economic and political impacts of the khat industry on agriculture, society, and people's livelihoods. A comprehensive study of khat also requires taking into account the social problems linked with specific misuses (which should be distinguished from the traditional uses) of khat and the resultant difficulties in the workplace and the home. In short, we need a collection of up-to-date studies on the production, trade, and consumption of khat from different regions of Ethiopia.

The purpose of this volume is to address the knowledge gap that exists in the areas identified above and to encourage evidence-based policymaking. The chapters deal with the expansion of khat consumption and the resulting growth of production and markets in Ethiopia. In many ways, the current volume expands the scope and builds on the themes developed in my book, *Leaf of Allah: Khat and the Transformation of Agriculture in Eastern Harerge, Ethiopia, 1875-1991*, which focused on Harerge and on the story of khat's transformation from an incidental shrub grown in small household gardens for the purpose of occasional consumption on cultural and religious occasions into the most lucrative cash crop in Harerge. Written by experts from various disciplines, the chapters in the current volume focus on various parts of Ethiopia, particularly those areas where khat is consumed and has become the cash crop of choice. Thematically, they evaluate both the impact of the commoditized cash crop on the age-old annual cereal complex that has characterized agriculture in many regions of Ethiopia for centuries and also look into larger processes of change that have led to the recent popularity of khat chewing, initiated agrarian transformations in different parts of Ethiopia, and improved rural life in several areas of that country.

In the second chapter, Hussein Ahmed, a historian, shows that the use of khat has been a longstanding practice among Muslim communities in Wallo. We do not have conclusive evidence as to how the practice of chewing came to Wallo. While there are some indications that the chew culture had been long associated with Yemeni communities in Wallo, the many associations with the *wadaja* ceremony Ahmed discusses point to a strong connection with the same institution in Harer, indicating that the chew culture likely came to Wallo from Harer (cf. Tesfaye, 1957, pp. 37-39; Jaenen, 1956, p. 184).

The third chapter focuses on the southwestern part of the country, Jimma, where khat chewing also has a long history. Focusing on what chewing means for Jimma's youth, Daniel Mains, an anthropologist, takes us into contemporary urban youth subculture and the way in which khat chewing helps the young cope with the difficulties of chronic unemployment, a condition that has little likelihood of abating any time soon. His interpretation of contesting narratives of khat's impact on youth subculture he collected in conversations with young chewers in the town shows a dialectical relationship between khat chewing and joblessness. By contrasting the experience of unemployed youth with those of young entrepreneurs, Mains describes the khat experience as a social milieu in which youth express their dreams, aspirations, and hope for the future. Together, both chapters provide contrasting images of the chew culture representing the use and misuse of khat.

My own chapter on consumption brings into focus the experiences that Ahmed and Mains describe. The chapter explains the reasons behind the explosion of khat consumption in major urban areas in Ethiopia since the late 1970s and argues that, contrary to the widespread perception that khat chewing is uniform, there are at least two distinct chew cultures. One is practiced mainly in rural areas – where chewing is governed by an evolved tradition, the other in urban centers where chewing is the primary setting for "killing time," an expression widely used to refer to a pastime activity. Overall, the chapter describes observations on the meaning of chewing in both the traditional setting and in the "modern" setting, highlighting differences between the two and challenging the persistent public perception of the chew culture that lumps together all forms of khat use. To make progress in reducing the undesirable aspects of khat chewing, the chapter argues for recognizing the contribution that traditional chewing could make in limiting the amount of khat that is chewed and for evidence-based debate rather than the gimmickry and demagogy that has characterized the public discourse so far.

With the fifth chapter, the focus of the volume shifts to the economic aspects of khat. My chapter deals with the production and trade of khat in Harerge, the area where khat culture has been dominant and khat's commodification began. It also documents the evolution of khat as an integral part of the agrarian system and its impact on household income, food security, and the macroeconomy.

In Chapter 6, Degol Hailu, an economist, examines the situation in the Southern Nations, Nationalities, and Peoples Region (SNNPR), which is the main supplier of khat to both the Addis Ababa and Jimma markets, and also in the Amhara regional state, which supplies a growing chew culture in Bahir Dar, the capital of the region. Focusing on two regions not known as centers of khat culture until recently, Hailu's chapter provides important data on the level of production of khat in the two regions and the significance of the revenue collected from khat for the regional economies.

In Chapter 7, Habtemariam Kassa, an agricultural economist, looks into major changes observed in the agricultural systems of the Harer highlands (the eastern section of the Harerge highlands) of Ethiopia, with special emphasis on the shift towards a khat-based farming system and regional economy. Based on studies undertaken over two decades, the chapter explains why the agricultural system increasingly shifted towards trees and shrubs, particularly khat. The chapter shows that the shift promoted crop-livestock integration, significantly improved farm income, and changed the economic profile of the region. Setting aside the fact that critics decry the cultivation of khat and its consequences, Kassa argues that the shift towards khat was unavoidable, given the absence of other feasible economic alternatives. While the author expresses concern about the long-term sustainability of the system, he predicts that khat production and marketing will remain important to the livelihood systems of the people in the Harer highlands and to the regional and national economies as well.

In the next chapter, Tesfaye Lemma Tefera and David Start assess the role khat plays in the livelihoods of a growing number of households and in the local economy of Deder and Meta districts in the Harer highlands with a view to informing the wider policy debate. The chapter shows that farming households, particularly those with access to irrigation, use the income from khat sales to finance the intensification of food crop production using improved seeds and inorganic fertilizers, thus successfully addressing their grain deficit. The chapter also shows the multiplier effects of recycled khat money when khat farmers spend their income locally on goods and services offered up for sale by others in the villages.

Chapter 9 directly addresses policy issues. It highlights the economic significance of khat to the farmers who grow it, the traders who market it, and the thousands of service providers who derive their livelihood from facilitating the delivery of khat from the growers to the consumers. While suggesting that, given its socioeconomic significance, a ban on khat would be ill advised the chapter also acknowledges that khat is not the ultimate panacea for the problems of Ethiopia's agricultural economy.

In the Afterword, Christopher Clapham reflects on the contribution of the chapters in this volume in the context of Ethiopian development policies and the place of khat production in the agricultural practices of the country. He views the increasing shift to khat cultivation as a reflection of the innovative capacities of Ethiopian farmers who for so long have been depicted as resisters rather than agents of change. After discussing the unique characteristics of khat that have enabled farmers to escape tight government controls, Clapham argues that neither international nor national prohibition of khat is achievable or even desirable since it is impractical and harmful to the millions of people who depend on it for their livelihood. In this regard, by cautioning that khat as an agricultural commodity cannot be expected to promote sustainable Ethiopian agriculture, Clapham accentuates the very message that the volume is intended to convey.

The immediate danger to Ethiopia's khat industry is posed by the threat of international prohibition or the cascading effects of national policies banning khat use and trade. The European countries that have determined that khat is not quite the socially harmful drug many opponents portray it to be might yield to pressure from the United States and Canada and outlaw khat. An international ban would have a devastating effect on the millions of Ethiopian farmers who grow khat as their primary cash crop. The long-term problem is the dilemma regarding the sustainability of household income and ecological balance within the context of an agricultural system based solely on the cultivation of khat.

PART I

CULTURE OF CONSUMPTION

Chapter 2

Tradition and Innovation in the Ritual of Khat Consumption in Wallo, Northern Ethiopia

Hussein Ahmed

෨෬

The study of the ritual of khat consumption in the rural and Muslim parts of Wallo and in the adjoining areas is important for a number of reasons. Firstly, it demonstrates the symbiotic relationship between pre or non-Islamic culture and Islamic informal acts of devotion. Secondly, it sheds light on the pervasive influence of Oromo culture as reflected in the language of supplication and titles used for, and by, the ritual officials of khat chewing sessions. Thirdly, it suggests that the ritual originated in situations where modern medical facilities did not either exist or were in short supply, or in which traditional therapy such as the use of herbs had failed to bring about speedy recovery from afflictions. The khat ritual therefore originally served as a last resort for solving problems affecting ordinary people. Fourthly, khat consumption was tolerated by the local *'ulema'* some of whom used the ritual as a forum for the teaching and propagation of a rigorous form of Islam, and for exhorting people to abandon popular ideas and practices they considered incompatible with the Qur'an and Islamic law. At the same time, they saw the khat session as an occasion for the enhancement of their religious status and a means for securing material rewards and recognition from the well-to-do and pious members of the community.

The bulk of the existing literature on khat (*catha edulis*; *qat* in Arabic; *chat* in Amharic; also called *meru* or *mira* in eastern Africa) contains only references to widespread cultivation in central and eastern Ethiopia. There is little discussion of the rituals associated with and dominated by the conspicuous consumption of the plant, which is a major theme of oral traditions.[1] In Yemen khat is known as *waraq al-habashī* (the Abyssinian leaf) or *waraq al-janna* (leaf of Paradise). The earliest written sources on khat consumption in Ethiopia are the chronicle of the mediaeval warrior-king, Amda Seyon (r.1314-1344) and the Arabic account of the sixteenth-century campaigns of *Imam* Ahmad b. Ibrahīm or Grañ (Marrassini, 1993, p. 54; Stenhouse, 2003, pp. 37, 104, 157, 159, 238). The khat plant originally grew wild but was later domesticated and cultivated in the warmer lowlands and in the eastern fringes of southeast Tegray, Wallo and Shawa, especially in several districts in Yajju, Ambassal, Qallu and Ifat. The widespread cultivation and marketing of different varieties of khat in Wallo is popularly attributed to the Yemeni emigrants who settled in north and central Ethiopia beginning in the late nineteenth century. They were already familiar with the culture of khat consumption and with the elaborate rituals connected to it in their homeland. This chapter examines changes in the pattern of behavior and symbolism in the ritual of khat chewing in Wallo.

TRADITIONAL KHAT CHEWING RITUAL

Long before its cultivation and phenomenal rise and expansion as a cash crop and an item of daily consumption in Ethiopia, often in competition with other food plants, khat was consumed, especially in the Muslim parts of Wallo, only on social occasions and religious festivals and by elders, and initially, by adult males. Specifically, khat was used at the *wadaja* ceremony at which elders and religious leaders gathered to offer collective prayers of supplication. Nineteenth-century European visitors to Wallo confirmed its widespread practice and rightly considered it an Oromo ritual (Isenberge & Krapf, 1843, pp. 323-324; Ficquet, 2003, p. 87; Gebissa, 2004, p. 10). It was later adopted by the Amhara of the region who became associated with the use of khat after their conversion to Islam.

The *wadaja* ritual in Wallo has at least two major types: *ya wand wadaja* (attended by men) and *ya dubarti wadaja* (attended by women), with further sub-types: for example, *ya tolfanna wadaja* (usually organized by childless families in which, instead of khat, *talla,* a traditionally-brewed ale, is consumed by the participants); *ya lej wadaja* (attended by young boys), about which the following couplet was composed: "*ya lej wadaja yawatal haja*" ("the *wadaja* of children leads to the fulfillment of wishes").

Broadly speaking, there are two categories of the ritual depending on the purpose for which it is held. The first, and the most prevalent, is the one solicited and hosted by an individual in distress or someone with specific problems, physical, material or mental. The second type is a communal one and is organized by the elders of the community in times of crisis such as famine, epidemic, war, or locust invasion. It is a time-honored tradition and a collective response to, and a mechanism for dealing with, a major crisis. The inhabitants of the village contribute materially and offer voluntary labor for the preparation of the ritual.

The personally-initiated *wadaja* is the most common one. A person in adversity announces his intention of hosting the ritual and, a few days or so before the date fixed by himself, he invites the elderly members of the community, especially those with long experience and reputation for proven and efficacious power of supplication, as expressed in the Amharic sentence: "*ekale yadar-*

ragallat wadaja maqbūl naw, aysetem" ("A *wadaja* officiated by so-and-so is accepted [by God] and never misses its target.")

The *wadaja* is scheduled to begin from either the early afternoon or the early evening, the latter lasting throughout the night until just before daybreak. Sometimes it lasts for several days, especially when, for example, a person is critically ill and does not recover quickly, and during *Pagumen* (locally called *Qwagme*), the five or six day long month preceding the Ethiopian New Year. The host also decides whom to invite, taking care not to exclude inadvertently famous people called *du'a' adragiwoch*, some of whom had wide reputation and were patronized by regional chiefs or even members of the royal family who possessed extensive land in Wallo like the late Crown Prince and Duke of Wallo Asfa Wossen Haile Sellassie,[2] who granted plots to the ritual leaders as a reward for their services (Ahmed, 2005, pp. 241-250). They prayed for the long life, prosperity and power of the local notables or members of the imperial family.

Before the ritual begins, the participants form a semi-circle around the leader who usually sits in the middle, flanked on both sides by men who take their seats according to their age. There is a partition (usually a thick piece of cloth) that separates the participants from the host who either sits down or is in a standing position behind the partition. The *jabana* (coffeepot), *sini* or *finjal* (cups), *rakabot* (a low wooden stand on which the cups are placed), and *gachcha* (a clay incense-burner) are put in front of the host.

The prospective participants do not demand or expect any material or financial reward for their services, although they accept presents which were offered to them voluntarily. Sometimes they provided basic items of consumption, especially if the host was too poor to afford them.

Preparations for the *wadaja* start a few days before the event. The amount and variety of the items for consumption depend on the financial position of the host, although some who find themselves in a desperate situation, spend so much that the expenses are a drain on their limited financial resources. Whatever the size of the *wadaja* in terms of the number of people attending it or the diversity and quantity of the items of consumption (principally coffee and tea), the one conspicuous item that should be available in sufficient quantity under any circumstances is khat.

Traditionally, khat was provided by a neighbour who had cultivated it in his private garden. In the nineteenth and twentieth centuries, before the cultivation and marketing of khat was diversified, in the Wallo countryside, there was a type called *jafjaf* with very coarse leaves. The person assigned to fetch the fresh leaves of the khat was called *awracha*. According to mystics, khat is seen as a sacred plant for the pious. The *awracha* is expected to be in a state of ritual or physical cleanliness (*tahara*) while cutting down the leaves from the khat tree. This rule is strictly observed.

In the course of the ritual the man in charge of distributing a fistful of khat leaves to each person is known as the *abba gar* (a phrase of Oromo provenance) or *barrak bay* (indicating possible Yemeni connection). He is considered as the principal officiator or manager of the ritual. After the first round of khat-chewing, people who want more khat rub the palms of their two hands together. This draws the attention of the *abba gar* to the request who places more khat into the outstretched hands of the participant who kisses the right hand of the *abba gar* as he receives the khat.

Soon after the main meal, the host offers the first round of coffee,[3] reciting the *basmala* (the Arabic acronym of *Bismi Allah al-Rahman al-Rahīm*) and formally offering it to the participants and asking for their blessings (Pankhurst, 1997, pp. 516-539; Ficquet, 2006, pp. 231-241). He says: "*awwal jaba, qaha jaba* [*qaha* is a corrupt form of the Arabic *qahwa*: coffee]: "[I] offer the first [round of] coffee", to which the *abba gar* responds by saying: "*hayat jaba, tena jaba, yawalladah, yawaddadah jannat yegba*" ("I wish you good life and health; may your parents and those who love you enter Heaven"). This is followed by a benediction, calling upon God to grant all the wishes of the host. After invoking the names of saints and appealing for their intercession and succor (by using the word *majan*), the *abba gar* invites the others to offer their own words of blessing to the host. He then asks them to forgive one another for past wrongs or misdeeds by saying: "*afu tababalu, gomi atyazu*": "say '*afw* [forgive] to each other; do not hold a grudge [against each other]."[4]

The khat ritual proper begins with the offer of a large bundle of khat by the *bala wadaja* or host with the following words: *chat jaba*,[5] *qanbat jaba* and the *abba gar* profusely bestows his blessings upon him. A repeat of the blessings is made by others up to three times. The host is then formally asked to state the purpose for which he has organized the *wadaja*. Complete silence reigns for a short interval after the khat has been distributed to everybody. Then the *qimaha* or the act of chewing (*bartcha* in eastern Ethiopia) begins in earnest (Ahmed, 1988, pp. 185-197).

When those assembled enter a state of ecstasy and euphoria (*wadajaw siggagal*: lit. when the *wadaja* gets hot or the participants feel 'high'), they continuously offer blessings to the host. The *abba gar* then narrates anecdotes about earlier *wadaja*s, the lives of locally famous saints, and the fate of hosts who had had doubts (*shakk*) about the piety of the participants or the efficacy of their prayers. Some Amharic poems in praise of saints are also recited, all of which add to the intensity of the devotional exercises. Coffee is served several times. First, the host says: *fere jaba, fere jaba*: (lit. May I offer beans [of coffee]). Then the pan on which the coffee beans have been roasted is passed around among the participants so that they can inhale the aroma. Depending on the financial status of the host, a sheep or a goat is also offered to receive blessings before it is slaughtered and the meat cooked and served at the end of the ritual. The host says: "*korma* (lit.: heifer) *jaba*" three times.

The *wadaja* reaches a climax when all attention is focused on listening to the exhortations and blessings made by the ritual leader. The whole session finally comes to an end with the final benedictions and supplications for the fulfillment of the wishes of the host (sometimes called *kaddam*; Arabic: *khadim*). Throughout the session there is a continuous supply of khat since nothing is more embarrassing to the host than being unable to provide it immediately upon request. If, however, the supply runs out towards the end of the *wadaja*, the *abba gar* says: "*chatu endallaqa azachen hulla yelaq*" ("As the khat is finished, may all our troubles come to an end"). In fact the host does not offer all the khat that he has fetched at once; instead he brings portions at regular intervals, the aim of which is to get the blessings of the participants as many times as possible, not to economize. Sometimes the leftover of the khat is boiled and the infusion served like coffee or tea. Such a drink is known as *hawza* and is believed to heighten the sense of stimulation known as *mirqana*, a term widely employed in khat-consuming areas throughout Ethiopia (Gebissa, 2004, p. 8).[6] Highly-stimulating varieties of khat are usually preferred.

The host often makes a pledge to prepare a feast if his wishes are fulfilled. Such a feast is called *barsa sadi* and the participants look forward to attending it since it proves that their prayers had been answered by God and enhances their reputation as efficacious intercessors. However, if the outcome of their spiritual exertions is contrary to their expectations and hopes (if, for example, a sick man dies or a bad situation persists), it is either explained away as God's will or recognized as though the ritual had not been performed properly, or the host had offended a participant, thus incurring his displeasure and becoming a victim of his wrath, as reflected in the expression: *qalb agaññaw* (he has been smitten by the 'heart').

Women are not excluded from the generally male-dominated *wadaja* as a matter of principle, but they gather either in a partitioned off space from the men or in an adjoining room. There are occasions when the hostess, usually the wife, organizes a *wadaja* to be attended by elderly women who are renowned for their piety and spirituality. The main characteristics of a women's *wadaja* are the absence of men and absolute concentration of the participants on the act of supplication, and their concern with matters related to the welfare of the womenfolk: recovery from illness, marriage of girls, childlessness and safe delivery for a pregnant woman. The *wadaja* performed for the last need to be met is known as *sittina* or *fatima qori*.

Two additional features of female *wadaja* that distinguish it from that of men are the preparation and consumption of *ganfo* (porridge) before and after the ritual, and the distinct formulae of supplication and blessings which are collectively and continuously recited in unison. The principal woman ritual leader says: "*rufo marabba*" [lit.: canopy of harmony], to which the others respond: "*marabba*".[7] After pronouncing the benediction, she asks: "*mache?*" [When?], and they answer: "*ahun*" [now]. Then she makes a long solo invocation to which both the hostess and the participants listen attentively, with the occasional expression of appreciation of her words or admiration for her attractive and enchanting voice.

The khat ritual has traditionally played a significant role in Wallo since it brought people together to participate in an act of communal worship and encouraged them to show concern for the well-being of others. It has provided an occasion for initiation into the world of spirituality. Through exhortation and the narration of anecdotes about the lives of saints, the people attending the ritual were introduced into the basic ideas of Islamic doctrine and practices.

Not everyone endorsed the role of khat in society. In the nineteenth century, there were heated debates among some Wallo Muslim scholars over the legality of khat consumption.[8] A few condemned it for three reasons: firstly, the practice was popularly and excessively venerated; secondly, the khat ritual leaders often claimed to have the power of intercession and mediation and of granting the wishes of those who sought their assistance; and thirdly, the leaders were accused of taking advantage of the credulity and desperation of the country folk and seeking material benefits to enhance their economic and social status. Moreover, khat sessions were also believed to lead to lethargy and distract people from their work and the performance of the obligatory daily prayers at the prescribed times. Others tolerated it arguing that the plant itself was so trivial that it was spat out in contempt after chewing it, and the mouth thoroughly cleaned (that is, long before it started to be washed down with water; in Harar this is known as *lullu qachcha*).

After the end of the Italian occupation in 1941, the rapid expansion of a cash economy, the increasing demand for consumer goods, the rising cost of living and growing unemployment have led to migration of people from the rural areas of Wallo to towns in search of new opportunities. The elderly men and women, especially those who possessed certain skills and experience while in the countryside, and were literate, were employed as Qur'an teachers in mosques and private houses. Some became traditional healers and *wadaja* leaders. Since the latter had no regular income, they gradually expected financial remuneration for their services, the amount of which varied according to the resources of the hosts. The khat chewing ritual has been affected by socioeconomic and cultural changes experienced by Ethiopian society since the Second World War.

CHANGES IN THE KHAT SUBCULTURE

A significant post-war development in Wallo and the rest of Ethiopia is the transformation of khat from an important ritual item into an agricultural commodity competing with coffee which it gradually superseded in terms of the area under cultivation and market value. This was an outcome of internal and external factors. As already noted, traditionally khat was supplied free of charge by those who cultivated it in their private gardens to the organizers of *wadaja*. In all likelihood, the plant had at first grown wild. Occasionally, however, whenever the number of participants in the khat ritual increased, especially in times of distress, its organizer probably offered to pay the khat cultivator for an additional supply. This must have contributed to increasing the volume of khat that was bought and sold in the market. It also became an incentive for its commercial cultivation.

Another important and perhaps more decisive factor for the commoditization and widespread consumption of khat was the arrival and permanent settlement of Yemeni immigrants both before and after the Italian invasion of Ethiopia (Ahmed, 1997, pp. 339-348; Gebissa, 2004, p. 82ff; Pankhurst, 2002, pp. 393-419; Anderson et al., 2007, p. 107). The Yemenis were already accustomed to growing and chewing the plant in their own homeland.[9] They were initially itinerant traders who visited khat-growing areas in the course of their travel but who later settled down in the main towns throughout the country where they eventually opened shops, teahouses and restaurants, and eventually emerged as wholesale and retail merchants. Some rented plots of land on which they grew khat for sale in open stalls in the towns;[10] but the majority bought the khat directly from cultivators and suppliers coming from the countryside or from the local khat traders. Since they had ready cash and were willing to pay more than what others offered, they became the favourite customers of the khat vendors (Gebissa, 2004, p. 94).[11]

The expansion of the trade in khat had important consequences for the local community and economy. Firstly, it gave impetus to the experimentation, acclimatization and cultivation of different varieties of khat. The *jafjaf* was replaced by several types with smaller and tenderer leaves whose potency was appreciated by the Yemenis, particularly by the older members who could not chew the coarser leaves because of bad teeth. So some had the tough leaves pounded in a mortar. The names of these new varieties reflect the influence of the Yemeni Arabs: *nashif, ratib, bustanī*, etc. Other varieties were cultivated later. In Dase, the provincial capital, and the surrounding districts, the type known as *galessa* (lit. strong) is popular with the well-to-do; naturally it is very expensive, particularly in seasons of short supply. In Kombolcha and its environs in northern Wallo, khat grown in such localities as Ergoyye, Qattataye, Qarsa, Warrabbacho, Garba and Dagan gradually became extremely popular. A small bundle of khat leaves tied together and covered with *embwacho* (*Rumex nervosus*) or fresh grass was known as *aqara*; a large one is referred to as *zurba* (Gebissa, 2004, p. 8). In recent decades, it has become usual for the tender leaves to be cut and put in plastic bags and sold to customers; these have acquired the name *ferfer*.[12]

The increase of the khat trade both in terms of demand, supply and consumption, and in the number of people earning their living from it (growers, vendors, distributors and transporters), and in the amount of land under cultivation – all at the expense of other essential food crops and coffee, its rival as a foreign-exchange earner – have had both salutary and damaging effects. For those directly engaged in cultivating, transporting and selling it, khat has led to unprecedented prosperity and a spectacular rise in their standard of living as they came to acquire wealth derived from profits earned from khat sales, and invested in other forms of property such as trucks and houses. The profit margin has been consistently high except when delay in delivering fresh supplies has suppressed it. The increasingly high rate of taxes, attempts to evade payment, and confiscation at customs posts has at times had a negative impact upon the trade.

Two recent developments connected with khat are the provision by a number of Muslim restaurants of special rooms for consumers (complete with mattresses, pillows and other amenities), especially for those coming from the provinces and the neighboring countries like Djibouti, Yemen and Saudi Arabia. They also provide what is locally called *shisha* (narghile) for the use of which clients are charged. The habit of smoking *shisha* has become popular (especially among girls and young women), and has spread into private houses and in the provinces. Cold drinks and tea/coffee are also served in the *shisha* houses [Arabic: *al-maqahī*] (Shash, 2001, pp. 2, 5).

The ubiquitous khat has had a more visible impact on the urban cultural landscape than is conventionally thought. Its consumption is no longer the preserve of Muslims only. It also has a significant clientele among Christians. It is not unusual to see the adherents of Islam and Christianity jointly attending khat sessions thereby sharing and experiencing the euphoria and sense of heightened consciousness induced by the chewing of the leaves, and contributing to interaction and a sense of solidarity between the two communities. Thus the khat ritual plays an integrative and cohesive social role in a country of ethnic, linguistic and religious diversity which is perhaps the most important characteristic feature of Ethiopian society.

REACTION TO PLEASURE CHEWING

Khat consumption in urban centers is said to have severe economic, psychological and social costs. Critics maintain that it contributes to a rising crime rate and loss of interest in work. Alcoholism is reportedly another negative byproduct of chewing. Regular chewers engage in drinking alcohol after the chew session, owing to the widespread notion that the effects of over-consumption (like loss of appetite and sleeplessness) can be mitigated by indulging in alcoholic drinks. This is popularly called *chabsi* (Gebissa, 2004, p. 9). For strict and practising Muslims, this is a flagrant violation of the Quran's ban on the consumption of hard liquor and an affront to Islamic morality and orthodoxy which ultimately undermines the social standing of the individuals concerned.

The issue of the permissibility of khat consumption has engaged the attention of ordinary and educated Muslims. Khat's pious defenders emphasize its sobering and inspiring effect, especially its power of inducing alertness and concentration, and creating a state of mind conducive to the pursuit of devotional activities. The merit of khat is expressed in a popular saying: *"yaqama tataqqama"* ("whoever chews stands to benefit."). Editorials which were highly critical of habitual chewing with some calling for its total and immediate ban or prohibition of the youth from indulging in it, have appeared in both Islamic and non-Islamic newspapers and other media, both private and public (*al-'Alam*, 1975; *al-'Alam*, 1976). Half-hearted attempts at curbing khat distribution and consumption without addressing the root-causes for habituation such as rising unemployment, and without taking into consideration the socio-economic implications of the ban, will likely remain unsuccessful in the foreseeable future.[13] As noted by Anderson et al. (2007): "The foreign exchange and tax revenues obtained from khat in Ethiopia have reached such proportions that reversing its cultivation is no longer an issue of plant extermination or alternative livelihood projects, but a question of how the development of the country will be financed in the absence of the khat industry" (p. 42).

Discussions of the impact of khat consumption have also been held in meetings organized by religious and public organizations. Government officials have made pronouncements against khat during the Derg period (1974-91). Khat was considered a harmful drug, the consumption of which diverted the young from revolutionary and political commitment and economic productivity, and even inspired conspiracy against the regime.

In the early 2000s, the Rūh Anti-Khat Club launched a campaign against habituation and posted Amharic slogans on the walls of buildings along the road running from Takla Haymanot Square towards the red pepper market in Addis Ababa. The following are some of its slogans: "Addiction to khat is a burden on the country and compatriots;" "Instead of chewing khat, let us nurture an anti-khat generation." "Use your time properly [and productively]; do not let your mind rust by chewing khat." The debate goes on, but it does not seem to have any impact in slowing down the spread of khat consumption and production.

NOTES

1. The literature on khat and its many related and relevant aspects is quite extensive. The present writer plans to compile a general bibliography on the subject in the not-too-distant future. The latest work is David et al. (2007). For a recent, pioneering and stimulating (!) study of khat cultivation and trade, see Gebissa (2004). See also his earlier contributions (Gebissa, 1997; Gebissa, 2003). See

also Carmichael (2000). For an earlier study, see Mercier (1980-1982).

2. Some of those who rendered such spiritual services to the local rulers received *waqf*-lands. See Ahmed (2005).
3. On the traditional coffee ceremony, see Pankhurst, R. (1997) and Ficquet (2006).
4. Note that the word *'afw* is Arabic while *gomi* is Oromo.
5. From which the word *jabata* is derived. *Jabata* is an article offered by someone (called *jaba bay*) seeking blessings, or the act of offering itself.
6. *Mirqaana* in Harar; defined as "the desired state of euphoria."
7. In the evening program of Thursday, 29 April 2004, the national television screened a program on Christian women's ritual in the course of which they repeated the formula of invocation similar to the one recited by Muslim women, but did not consume khat.
8. A long section on the issue is included in the Arabic hagiographical work cited in Ahmed (1988).
9. On the Yemenis, see Ahmed (1997) and Gebissa (2004). On relations between Ethiopia and Yemen, see Pankhurst (2002). In Kenya the Yemenis of Hadramawt promoted the consumption of the plant. See Andersen et al., 2007.
10. A famous retailer of khat in Dase was the late 'Abdallāh 'Amr al-Jawmarī.
11. Gebissa (2004) noted the principal khat exporters from Dire Dawa to Aden until the 1960s were Yemeni Arabs.
12. In the Addis Ababa *chat* market, the traditional varieties such as *walane* and *walqitte*, brought from the Gurage region, have in the last twenty years or so been superseded by *sabbata, awaday, galamso, wando, balache* etc. Most varieties were named after the places they were cultivated or the main centres of distribution. Moreover, the khat trade led to the opening of innumerable small shops in distict quarters of towns where the product was displayed openly for customers. Besides the old *chat* quarters concentrated

in the Mercato area of Addis Ababa, there are now khat-selling shops scattered throughout the city. That testifies to the spectacular increase in the quantity of *khat* brought for sale and in the number of habitual consumers. Some of the city's young and affluent residents spend the weekend chewing the plant along the roads leading to the capital's suburbs.

13. There is a story, both amusing and enlightening, about a law introduced in 1975 by Muhsin al-ʿAynī, the prime minister of what was then North Yemen, banning the use of khat which aroused a public uproar calling for his immediate resignation. Frightened by the outcry, the official duly resigned and was seen the following day taking part in a khat-chewing session together with his relatives and compatriots, including those who had demanded his resignation.

Chapter 3

Chewing and Dreaming: Youth, Imagination, and the Consumption of Khat in Jimma, Southwestern Ethiopia

Daniel Mains

ഩരു

"Khat kills time. Youth are simply sitting and chewing. Today's generation isn't interested in working; they are only interested in chewing." In conducting research on youth culture and unemployment in Jimma, Ethiopia, I was initially surprised by how often people answered my questions concerning work with statements like this about khat. For urban Ethiopians, the problem of unemployment is conceived as inextricable from the consumption of khat.[1]

In conversations and interviews with youth and adults two primary perspectives on the issues of khat and employment emerged. It is argued that khat either causes unemployment or is a result of unemployment. In the first perspective youth do not work because they prefer to chew khat. This perspective assumes that economic opportunity is present in Ethiopia, and that at the very least an enterprising young person can create his own work. The hours spent chewing khat every afternoon could be spent working, but either through "addiction" or personal choice youth pass their time with khat. In terms of policy, individuals who support this position argue that by banning or controlling the use of khat the government may decrease unemployment. The second perspective argues that khat use is a symptom of the wider problem of unemployment. From this perspective, youth want to work but there is simply nothing available. Without work or any other recreational options young men have nothing else to do except chew. From this perspective, in order to decrease the use of khat there must first be an increase in economic opportunity.

In this paper, I argue that both of these perspectives are correct but they are also oversimplified. The first perspective assumes that drug use causes the social condition, and the second argues that the social condition determines drug use. The interaction between the experience of khat and the condition of unemployment is not examined. A close exploration of the practices and narratives surrounding khat use among young men reveals the reciprocal nature of the relationship between khat use and unemployment. In discussing khat use, young chewers and non-chewers both speak about khat in terms of dreams, desires, and hopes for the future. In a social context of extremely limited economic opportunity, chewing khat provides a means of mediating the gap between one's desires and reality. In this sense, the experience of khat chewing is shaped by the social condition of urban poverty. At the same time, the particular practices associated with khat use among the unemployed contribute to the reproduction of their joblessness.

I begin this paper by describing the problem of unemployment for young men in urban Ethiopia. Young men represent a particular sub-culture of khat consumers and the practices and discourses that surround their khat use must be understood in relation to the struggle to attain their aspirations in a context of limited economic opportunity. Although khat is associated with a variety of different forms of social interaction including religious and healing ceremonies, convivial discussions, and celebrating holidays or weddings (Gebissa, 2004), my primary interest is in how urban young men chew. I examine the dreams and desires that emerge among unemployed young men through the process of chewing khat. I contrast the experience of unemployed youth with youth entrepreneurs to illustrate the importance of social context for the manner in which khat is experienced.

Debates over the consumption of khat in urban Ethiopia are rooted in contrasting conceptions of the relationship between youth and imagination. For young men khat is a vehicle for imagining creative solutions to problems associated with unemployment. To chew is to dream of alternate possibilities for the future. For those who are against the consumption of khat, the social category of youth is imagined in relation to khat consumption. On the one hand, khat is a tool that facilitates the consumer's imagination, and on the other, khat is a symbol that draws the imagination from others towards the consumer. Ultimately, I argue that the dual function of khat as tool and sign enables young men to negotiate their apparent lack of agency and inability to attain aspirations.

UNEMPLOYMENT AND YOUNG MEN'S PROBLEMS OF PROGRESS

Rates of unemployment among young people in urban Ethiopia are around 50 percent and lengths of unemployment average three to four years (Serneels, 2007). In this context, young men experience significant difficulties in attaining their aspirations. The aspirations of young men are structured around an abstract notion of progress. Young men explained to me that a good life is one that involves change or improvement over time. In contrast, unemployed young men often lamented the fact that their own lives do not change from day to day. On more than one occasion unemployed young men made comments to me like, "We are just living like chickens, eating and sleeping." Put plainly, a life that consists only of filling one's stomach and sleeping is thought to be meaningless.

To some extent, frustrated aspirations for progress are particular to the generation of youth that have come of age following the overthrow of the Marxist Derg regime in 1991. The young men I studied were far more embedded in an ideology of progress through education than their parents' generation. Most urban youth were the sons and daughters of parents who were raised in rural environments and did not advance beyond primary education. Despite living through a Marxist revolution that was associated with particular notions of modernity (Donham, 1999), their lack of education meant that the parents of the youth in my study often did not internalize an ideology of progress as it pertained to their own lives. It was common for parents of unemployed youth to explain generation differences in terms of education and desire. Adults claimed that their children had much greater levels of both. A similar perspective is voiced in this quote from a young man, "Today's generation is different. In the past, everyone expected to do the same work as their parents. Today everyone wants to learn and to have a better life. If someone's father is a farmer then he wants to be a modern farmer, or else to do a different job altogether." For both young people and their parents, education is linked with elevated aspirations for the future. In describing their life histories, most parents spoke of the movement from a rural area to Jimma as a major shift in their life. Upon arriving in Jimma, they generally accepted whatever work was available and were not as concerned with issues of status as their children. Parents often argued that their children's lives should be different from their own specifically because of their higher level of education, and they were disappointed when this was not the case.

The relationship between formal education and the production of progressive aspirations is complex. It is not simply the case of a modern institution producing a modern worldview.

The value of education in accessing desirable employment was largely due to the particular nature of traditional Ethiopian power relations, especially as they existed in the northern highlands. Education provided a means by which those power relations could be mapped onto an urban context in the form of a government bureaucracy (Hoben, 1970; Poluha, 2004). The ideological implications of formal secular education are perhaps less dependent on context. As a student, time is divided into neat units, and change is experienced as linear improvement. When one advances from grade to grade, it is assumed that this movement has created a change within one's self as well. The educated individual expects to be transformed so that his future will be better than the present. Unemployment is the absence of change. Days pass but one's material and social position remains the same. Long-term unemployment prevents youth from imagining a desirable future and placing their day-to-day lives within a narrative of progress. As one young man who had been unemployed for two years after completing grade 12 put it, "When I was a student I had no thoughts. I learned, I studied, and I did not worry about the future. Now I always think about the future. I do not know how long this condition will last. Maybe it will be the same year after year."

In urban Ethiopia, concerns about progress and the future are expressed in terms of the specifically local idiom of social relationships. In the progressive narratives of young men, one first completes education, followed by employment, economic independence, marriage, children, providing support for one's extended family, and helping one's community. Progress is conceptualized as linear changes in one's position within social relationships, characterized by the movement from dependence, to independence, to providing support for others. The difficulty of obtaining desirable employment after completing one's education has ruptured this narrative. In practice, young men are dependent on their parents for increasing lengths of time. Many of the young men in my study were in their late twenties or early thirties. They had been unemployed for four or five years and they continued to depend on their friends and family for daily handouts. These young men had little hope for attaining their progressive aspirations in the future. Unemployed young men argued that khat consumption was valuable precisely because of its potential to reposition oneself in relation to the future and reengage with a hopeful narrative of progress.

THE DYNAMICS OF KHAT CONSUMPTION AMONG UNEMPLOYED YOUNG MEN

Before analyzing the implications of khat for young men's imaginations of the future, I describe the day-to-day dynamics of khat consumption in Jimma. The town is widely believed to be one of the centers in Ethiopia for khat consumption. Many residents claim that after Harer and Dire Dawa, Jimma has the highest per capita rate of consumption. Most of the young men I spoke to estimated that 70-85 percent chew at least weekly, and half this amount chews daily. They stressed that khat use in Jimma varies significantly with neighborhood and religion. For example in a predominantly Christian neighborhood approximately 50 percent of youth may chew, while up to 90 percent of Muslim youth may chew khat. While these figures do not necessarily correspond with the actual number of young chewers, they certainly reveal a perception that khat consumption is very wide spread. A study of khat consumption, carried out among secondary students in Agaro (a smaller city to the north of Jimma) documented similar rates of consumption (Adugna et al., 1994).

Khat has traditionally played an important role for Ethiopian Muslims in prayer, celebrations, and community discussions (Gebissa, 2004). Although more research on the history of khat in the Jimma area is necessary, based on conversations with residents, khat consumption in Jimma appears to have followed patterns observed in other areas of Ethiopia. Like other areas, khat has been popular among adult Muslims for some time. Heavy khat use among Christians and youth emerged during the latter part of the twentieth century and continues to be largely confined to young men in urban areas. This means that urban youth can be understood as developing their own culture of khat consumption. This culture certainly draws on the practices of the past, especially the relationship between khat and discussions and problem solving. Many of the practices I will describe below, however, are innovations developed primarily by urban youth. Chewing khat while watching videos and drinking alcohol after a chew are two examples of urban youth innovations. As I detail below, the culture of khat consumption among urban youth is specifically related to a context shaped by rising unemployment and the devastation caused by HIV/AIDS.

The relationship between khat and ethnic and religious identity in Jimma is complex. In contrast to Daniel Varisco's (1986) discussion of khat consumption in Yemen, khat is not seen as an expression of an Ethiopian identity. The Ethiopian state has historically been dominated by Orthodox Christians, and for some individuals consuming khat is thought to signify a Muslim identity. However, young men in Jimma often claimed that their shared khat consumption was evidence that ethnic and religious divisions are not important for their generation.[2] I frequently observed young men of different ethnicities and religions chewing khat together during the extended period of mourning following the death of a friend. Often these diverse groups of young men shared funeral tents with older Muslim men whose Christian peers did not join them in chewing khat.

For urban young men there is little ethnic or religious variation in the practices of habitual chewers. Especially among the unemployed, the first half of the day is usually spent trying to scrape together a few birr for khat. The price of khat is relatively low in Jimma and at the time of my research (2003-05), two or three birr was usually enough for one person, although five birr was the preferred amount if finances were available. Depending on personal preference, khat chewers invest additional money in better quality khat or simply purchase a greater quantity. Money for khat may come from performing odd jobs like simple manual labor or watching a friend's store for a few hours. Borrowing or pilfering from one's parents is also very common. Most young consumers of khat have a few friends that they regularly chew with, and if one of them is able to buy khat, he will share with his close circle of friends. In general, despite being unemployed, one way or another, most habitual chewers manage to obtain khat on a daily basis.

For the unemployed, khat chewing usually starts in the early afternoon after lunch. Khat chewing is nearly always a social activity involving at least two or three friends. If one of the young men has access to a house where they can chew without being disturbed, this appears to be the ideal location. However, in many cases parents do not permit their children to chew at home. In Jimma, adults rarely venture into the types of houses described below and nearly always chew at a private residence.

In Jimma, khat houses are very common near any of the primary areas where khat is sold. These houses are filled with low benches or mats where men can relax, smoke cigarettes and chew their khat. A few houses allow chewers to smoke a hookah (*shisha*) for a small fee, but this is not common in Jimma. Coffee and tea are sold at a low price and people are free to sit and chew for as long as they like. The social dynamics of khat houses varied considerably in terms of the space that was present for interaction. At one house, that I spent time at young men lounged on mattresses chain smoking. Some young women were present and they were chewing as well, but in general it is not common for young women to chew in public and this was the only khat house I observed where they were present. Ja Rule and Eminem (popular American rap musicians) blasted from a small boom box, making conversation nearly impossible. Youth on the mattress leaned on each other sometimes joking with the person sitting next to them, but mostly staring off into space or singing along with the music. At another house that I frequented with one of my primary informants, the atmosphere was entirely different. This house was known as the "*Hoolaa* Café." It was explained to me that *Hoolaa* is the Oromo term for sheep and the name was given to the house because of its close proximity to the daily sheep market. The house was made from sheets of corrugated tin and had wooden benches instead of mattresses. At the *Hoolaa* Café the main attraction was conversation. This house attracted a crowd that was interested in talking about current events, and chewers would sometimes bring a newspaper with them in order to stimulate conversation. The customers were usually in their twenties and although they were all unemployed, they frequently had a year or two of post-secondary education. Where the clothing of the youth at the first house tended towards baggy jeans and sports jerseys, the young men at *Hoolaa* Café preferred the more conservative style of button down shirts tucked into pants. Although the customers at the two houses appear to be segregated roughly in terms of parental class background, both houses are in the same neighborhood and sell coffee at the same price.

Video houses are also a popular option for young khat chewers. Video houses are typically single rooms filled with rows of wooden benches that face a television. Beginning in the afternoon videos are shown. The films are usually from America or India, but the occasional Ethiopian film is shown as well. For the price of fifty *santims* (cents), video houses provide young, almost exclusively male, chewers with a comfortable space to escape from the heat and lose themselves in chewing and watching a film.

The choice of where one chews is clearly very important. From the loud music of the first house, to the discussions of current events at the *Hoolaa* Café, to intense engagement with films at the video house – in each case the experience of khat is mediated by the space in which it is consumed. Young men are highly selective about where they chew and regulars at the *Hoolaa* Café rarely venture into houses that do not provide a conducive environment for conversation. The self-segregation that chewers impose on themselves produces different social networks that may have implications for accessing economic opportunities. The relatively high level of education possessed by most of the chewers at the *Hoolaa* Café meant that they were interested primarily in government employment. Many of them also earned occasional incomes as brokers, bringing together buyers and sellers of rental houses, mobile phones, electronic equipment, or any other lucrative item. Chewers at the first house were generally less educated and more deeply immersed in the urban informal economy. Further research is necessary to determine if social networks that are generated through khat consumption have long-term economic implications for the reproduction or subversion of class-based stratification.

CHEWING, DREAMING, AND *MIRQANA*

For most youth the purpose of chewing khat is to reach *mirqana*, the high attained approximately an hour after beginning to chew. During *mirqana*, the heartbeat is notably faster, one begins to sweat, and there is a general sense of happiness and satisfaction with life. It is during *mirqana* that thoughts and conversations turn to the future, and youth begin to "dream."

Youth generally chew for at least three or four hours, finishing at some point in the evening, when the sensations associated with *mirqana* begin to shift, conversation slows down, and the chewer often becomes more introspective. It is at this point that many youth take *chabsii*, a period of drinking that is intended to "break" the *mirqana*. Those who drink *chabsii* often claim that it helps to eliminate the period of thoughtful introspection near the end of *mirqana* and facilitates sleep. In Jimma, the most common form of *chabsii* is *tej* (honey wine). Youth will frequently drink at least two or three *birillie* (a glass container shaped like an Erlenmeyer flask) of *tej* and then if finances are available go to a bar to sip a single beer and socialize with friends. Not all khat chewers take *chabsii* and some are very critical of the activity. They argue that khat in itself is not bad, but after drinking one is likely to engage in theft, irresponsible sex, fighting, and other potentially dangerous activities. Drinking is thought to be especially dangerous after khat because the amphetamine-like nature of the drug allows chewers to consume particularly large amounts of alcohol without feeling intoxicated. Young men who do not take *chabsii* usually return to their homes after chewing to read or watch television. My concern in this paper is not with the interactions and sensations associated with chabsii or the post-*mirqana* experience of non-drinkers. Rather my interest is in the social practices and discourses that are produced during *mirqana*.

It is during the state of *mirqana* that the particular social interactions associated with khat begin to emerge. Among unemployed young men, the state of *mirqana* is associated with intense discussions and "dreams" about one's desires for the future. Chewers describe the opening of the mind (*aimro yikeftal*) and thinking about wishes (*miññot*). In this paper, I use the term "dreaming" in order to describe the general process of talking and thinking about the future during *mirqana*. In Amharic youth use a variety of terms, including dream (*ḥilm*), goal (*alama*), wish (*miññot*), and hope (*tesfa*) to describe where the mind is directed during *mirqana*, but when speaking in English they usually use the word "dreaming." Among young men, this English word is frequently incorporated into Amharic.

If possible, young men prefer to chew with the same two or three friends with whom they are comfortable and who have similar interests and conversational styles. Individuals usually come in and out of the conversation, sometimes giving passionate monologues that last five to ten minutes, and at other times staring off into space, lost in one's own thoughts. For youth who watch videos while chewing, conversation is limited. In their case, the video appears to act as a foundation or stimulus that allows one's thoughts to travel outside the boundaries of normal life.

In the process of researching the consumption of khat among the unemployed, I spent numerous afternoons with small groups of young men as they chewed and talked. The following discussion of *mirqana* comes from a group discussion I conducted with Habtamu and two of his friends as they chewed khat. At the time of my research, Habtamu was in his mid-twenties and unemployed. Together with his friends Yonas and Tawfiq, he chewed khat almost daily in a shaded area of Yonas' spacious family compound.

Habtamu: During mirqana you *arrange*[3] your mind. Now, for example you might be thinking about doing some kind of work or some kind of study. That thing becomes very broad. Your mind thinks about it. Anything that you want to do, your mind is able to grasp it very quickly.

Daniel Mains: For example, if one person has a problem and he chews khat will he find a solution very quickly?

H: Yes. *Around the table*, a good *diplomat* will be created. This is the main thing about khat. When there is an argument about something for example if there is a problem between husband and wife among Muslims they will give the elders one sheep and a lot of khat. The elders will chew the khat and solve the problem.

Yonas: Yes. There is another thing. When you chew khat you think tomorrow I will buy a car. Tomorrow I will get money from the bank and buy a house.

Tawfiq: When you're chewing.

Yonas: But in the morning if you ask me, I have nothing.

H: Usually it is like this.

Yonas: It is mirqana.

On other occasions, I joined a young man named Alemu who was in his late twenties and had been without regular work during most of the nine years since he finished grade 12. He lived with his mother who owns a small shop, and he had been chewing khat daily since finishing secondary school. His nickname among his friends was Mulu Qen, meaning full day, because he would chew khat during the entire day. Alemu and I met in a small khat house where young men were seated on low benches or sprawled on a mattress, chain smoking, drinking coffee and chewing. Shouting over hip-hop blasting from the tape deck, Alemu explained that during *mirqana* his future seems "very bright" and all of his dreams are within reach. "When I chew khat I get happiness, if I think about New York City it is like I am actually living there."

The dreaming that is associated with khat consumption is not a passive activity. As the quotes above indicate, the social dynamics of khat consumption are very important. Chewers exchange ideas and plans for the future are formed through intensive dialogue. It is the process of "dreaming" that appears to bond khat chewers together as a social group. Alemu classified others in terms of chewers and non-chewers. Especially during *mirqana* he prefers to be around chewers. He described chewers as dreamers who are able to go places in their mind, while non-chewers are continually caught up in day-to-day affairs. During *mirqana* everyday reality seems very dull. Chewers do not want to talk about anything "normal" (the English word is often contrasted with dreaming). They prefer an atmosphere that allows them to escape into explorations of hopes and desires for the future. Alemu's favorite place to chew was the *Hoolaa* Café where conversations tended to focus on politics and business opportunities. The particular setting facilitates the generation of dreams that appeal to the individual khat consumer.

Many of the non-chewers that I spoke with were students at the Jimma Technical School. For the students, "dreaming" and discussions about seemingly unrealistic hopes for the future were cited as one of the key reasons for avoiding khat. They gave examples of the sort of wild fantasies that chewers entertain during *mirqana*. Chewers talk about owning seven airplanes or building a three-storey hotel. Non-chewers argued that this type of conversation is meaningless. The students are still in the process of pursuing their future goals through education. For them it is better to focus on more realistic goals that are close at hand and can be attained by passing through clearly defined steps of education and then work. From their perspective, sitting and talking about what is clearly impossible has no practical value.

Critiques of chewing and dreaming also made the point that all of the planning and discussing of the future that takes place during *mirqana* is lost the following morning. While khat stimulates the mind to form detailed plans for achieving even the most impossible desires, chewers cannot seem to remember these ideas once the *mirqana* has passed. Khat chewers also acknowledged this problem. In the group discussion with Habtamu and his friends quoted above, Habtamu described how one feels the morning after chewing khat.

> In the morning you don't want to talk with anyone. You won't even greet your friends. You try to remember what you were thinking about the day before, but it just won't come. Nothing helps, you can drink coffee or try anything, but you just feel bad. In the afternoon, after khat you can deal with people again and all of the plans from yesterday come flooding back. The only thing that will bring back the feeling of brightness and hope for the future is to chew more khat.

Most of the unemployed chewers spoke of an intense depression that occurs on the days when one does not obtain khat. Alemu simply stated that, "If I don't get khat, I hate everything." Both working and unemployed young men explained that the best part of their day was when *mirqana* was beginning to set in, and they dreaded those days when they could not scrape together the money for a small bundle of khat. Habtamu's friend Yonas who was also unemployed and in his mid-twenties explained that, without khat in the afternoon, the only way to pass the time is to wander around. The sun burns into his head and he begins to think about all of his problems - the lack of a job, friends who have died of AIDS, and the hopelessness of the future. He claimed that if he does not get khat he inevitably ends up spending a night or two in jail. I found this to be very surprising because Yonas had an extremely calm and polite demeanor. However, without khat he wandered around in a cloud of depression. Yonas explained that he would eventually run into an acquaintance and begin exchanging insults, which then escalate into a fight that will attract the attention of the police.[4] Only partially in jest, some of the technical school students argued that khat's ability to prevent depression and fights like this is the reason why the Ethiopian government does not ban it. It prevents people from thinking too much about the problems in life, and as long as they obtain their daily khat, they are unlikely to disturb the government.

Among unemployed youth, khat creates a cycle of dreaming and forgetting; satisfaction and depression; chewing and waiting to chew. In order better to understand how this process is intertwined with the condition of unemployment, it is necessary to look more closely at the specific types of dreams that emerge during *mirqana*. The nature of the dream is shaped by the economic conditions of urban Ethiopia. At the same time, it appears the process of chewing and dreaming facilitates the reproduction of those conditions. In the section that follows, I treat the dreams that emerge during khat chewing sessions among young men in Jimma as a discourse that sheds light on the particular relationship between khat and the context in which it is chewed.

DESIRE, DREAMS, AND EMPLOYMENT

In terms of timing, sex seems to be the most short-term of the dreams or desires associated with chewing khat. In discussing khat and sex my goal is not establish a causal relationship between khat and young men's sexual behavior. Rather, I wish to demonstrate that youth discussions of khat and sex are means of dealing with issues of unemployment, HIV/AIDS, and desires for the future. Young men claim that during *mirqana*, once a person begins thinking about sex it is nearly impossible to think about anything else. One unemployed young man described a process in which the chewer's desire is so intense that a woman may absorb his "heat" and increase her sexual appetite. Some young men claim that having sex after chewing khat is not so much a desire, but a need.[5]

In Jimma, young men often perceive khat as a drug that causes their sexual desire to escalate to a potentially uncontrollable level. Dreams concerning khat and sex are shaped by two important social conditions. One is the extremely high prevalence of HIV in urban Ethiopia. In describing the sorts of thoughts that occupy his mind when he does not get khat, one unemployed young man explained that in the past friends disappeared due to prison, travel abroad, and disease; today it is just disease. It appears that khat provides the ability to overcome much of the fear and depression surrounding HIV. As noted above, during *mirqana* one is filled with a sense of happiness and hope. Sorrow concerning lost friends and fears related to HIV and sexual intercourse temporarily disappear.

Unemployment is the other important factor conditioning the relationship between sexual desire and khat. In narratives from unemployed young men, sex is continually linked with money and employment status. Young men in Jimma argue that in Ethiopia there is no love; there is only money. They see this as a natural reaction to the condition of poverty in urban Ethiopia. Without work or access to money it is very difficult for a young man to fulfill his desire for sex. As one young chewer put it, "without money I must sit with my desire." At least temporarily, khat allows this problem to be overcome. The notion that the "heat" of desire that is produced during *mirqana* may be transferred to a woman is a good example of the sort of dream that occurs while chewing khat. Something that is made very difficult by economic constraints is imagined to not only be possible during *mirqana*, it is actually facilitated by the act of chewing khat.

Other statements from young men demonstrate the ambiguous nature of the relationship between khat, sex, and HIV. In a discussion with Habtamu and a group of friends regarding khat the conversation took a tangent towards women and relationships. These young men argued that relationships with women could lead to a number of problems including illegitimate children and HIV/AIDS. As usual, Habtamu articulated the group's ideas and argued that the solution to this problem was to chew khat. He explained:

> Usually if you don't chew you get these problems.
> If you don't chew and you are on the street, some

women will come, you greet them, invite them for coffee, *communication* is created, and a relationship begins. After that you will enter into bad problems. Therefore if you chew, you enjoy *mirqana*, you go home, you watch TV, you read, that's enough – you sleep. *Morning*, you get up at ten, you eat breakfast, you eat lunch, and then you chew khat.

Chewing khat removes one from the public space that enabled interactions with women that could lead to potentially serious consequences. As the conversation progressed, these young men soon adopted the exact opposite position, arguing that in fact khat does lead to irresponsible sex.

In fact, khat neither leads to, nor prevents particular sexual practices. Young men experience intense desires that are interrelated with issues such as unemployment and HIV/AIDS. Khat facilitates an exploration of these desires and the barriers to their fulfillment. Contrary to claims of adults and some youth, khat does not so much cause the spread of HIV/AIDS or unemployment as much as it provides space for constructing one's own position in relation to these issues. In discussions that occur during *mirqana*, an idea is presented (for example, khat leads to irresponsible sex) and then evaluated and critiqued. The conclusions that are reached in discussions fueled by khat are often forgotten, but the process itself is important. The simple act of discussion allows young men momentarily to come to terms with the gap between their desires and economic realities.

Conversations that take place while chewing are often repetitive and as many young men chew daily, what begins as fantasy is frequently perceived as a realistic possibility. Possibly even more common than discussions of women and sex are conversations about migration, especially to the United States. Through these fantasies, aspirations are developed in a manner that is intertwined with economic behavior. As noted above, a key problem that young men face is the inability to experience progress through time. The dreams associated with *mirqana* are continually directed towards the future and young men conceive of hopeful possibilities.

The following comes from the conversation between Habtamu, Yonas, and Tawfiq that I quoted above. They were describing the types of conversations that they have during *mirqana*. They were also experiencing *mirqana*, and this most likely influences the flow of the discussion.

> Tawfiq: If you are thinking about the *future* or another thing….
>
> Habtamu: *Future?!* In the name of God! Stop! [implying that this is something that he could talk about all day]
>
> Tawfiq: If you buy khat today, you will become bright. You will plan ten years ahead. [He names off one thing after another – work, marriage, children – snapping his fingers between each one.] Now if I am experiencing *mirqana* and you ask me I will tell you about America. After arriving in America I will go to Atlanta.
>
> H: Is it Atlanta today?
>
> Daniel Mains: My town.
>
> Tawfiq: Yes, yours. There, good work. A private car. A good house.
>
> Yonas: Palace.
>
> Tawfiq: After I win the DV [Diversity Visa Lottery], I will do all of this. But sometimes it gives me a headache. Thinking, "why can't I do this?"
>
> Yonas: Why is it this way? *America life* and *Ethiopia life* – there is one hundred years *difference*. An American will never think about finding something to drink or something to eat. For us we eat breakfast at home, after that we don't know where we will find lunch. This is a problem.

During *mirqana*, the relationship of young men to their future is shifted and the struggle to experience progress is no longer a source of anxiety. Introspective thought about one's future brings pleasure rather than stress, but the shift is not complete. As the comment about the headache indicates, the ability to achieve this fantasy is continually questioned. However, these doubts appear to be overcome by the patterned nature of the conversation. It is not only electronic media that facilitate the imagination of possible lives in the manner envisioned by Arjun Appadurai (1996). In this case, khat-fueled conversations generate imaginative futures. When Tawfiq mentions that he plans to go to Atlanta, Habtamu's response indicates that they have had this conversation before and the only difference is the name of the city.[6] The conversation is creative but within boundaries, and these boundaries shape the construction of the possible.

International migration, especially to America, was a constant topic of discussion among young men I knew in Jimma. Although this dream is not unique to khat chewers, it appears to become far more tangible during *mirqana*. Chewers discuss particular cities like Las Vegas or Atlanta and stories they have heard about life in America. To some extent narratives about migration are fantasies about a country that is primarily known through films and second or third hand stories.

Like discussions of sex, dreams of international migration are conditioned by unemployment. The most obvious problem associated with unemployment is a lack of money, and travel to America provides a clear solution to this problem. Many unemployed young men have been out of school for at least four years and they have been without regular work for much of this time. As they grow older and continue living with their parents and chewing khat everyday, they face an increasing loss of respect from their families and neighbors. Young men describe frequent conflicts with their parents who criticize their behavior and accuse them of being *duriye*[7] (juvenile delinquents). Despite the fact that many young men are reaching the age of 30, there is no visible end to their current state of idle chewing.

During *mirqana,* a solution to this problem is not only visible, but it appears to be easily attainable. Non-chewers sometimes described their friends who have developed serious khat habits as people who for one reason or another have lost hope. In this context it seems as if habitual chewers have an addiction to the sensation of hope that is only experienced during *mirqana.* Khat enables young men to construct a vision of the future in which they are able to experience the progress they desire.

The experience of khat among young men in Jimma is specifically shaped by the condition of unemployment. *Mirqana* brings a general feeling of happiness and forgetting of one's problems that seems to be associated with many recreational drugs. However, the narratives of youth reveal that khat is chewed not only to escape the present, but also to plan and dream about the future. Dreams of travel abroad or starting a small business allow youth to envision a future in which they have money, respect, and independence from their parents. In this sense, it is not necessarily khat that causes young men to become idle dreamers. Long-term unemployment provides an environment where youth need to dream in order to escape the hopelessness of their situation. This dynamic is not limited to Ethiopia. In the context of economic decline and widespread unemployment, produced in part by neoliberal policies of structural adjustment, dreaming about international migration has become an important practice for urban young men across Africa (Gondola, 1999; Newell, 2005). In the Ethiopian case, khat simply provides a means for young men to deal with the excess time and bleak future that characterize the condition of unemployment.

If khat helps solve problems associated with unemployment it also makes it very difficult for young men to find work. The simple fact that khat consumes an enormous amount of time prevents many young men from doing anything other than sitting and chewing for at least half of the day. A number of chewers told stories about seeing postings for available jobs but choosing to chew khat with friends instead of pursuing the position. The cycle of dreaming and forgetting makes it very difficult for youth to put their plans for the future into action. In those moments when they are not under the influence of khat young men are frequently too depressed to carry out the plans that were developed during *mirqana*. As Habtamu put it during a morning of sober of reflection, "With khat I can forget all of my problems. I do the same thing that I have been doing everyday for years. If I did not chew, I could not tolerate this situation day after day. Without khat I would have to do something to change my life." The process of chewing khat among the unemployed also tends to be disempowering in the sense that the particular dreams that are generated during *mirqana* are outside of the control of the chewer. For example, outside of entering the Diversity Visa Lottery every year, there is a little that an individual can do to improve his chances of traveling abroad. Sitting, waiting, and chewing appear to be the only possible actions that a young man may take.

CHEWING AND WORKING

The close reciprocal relationship between the experience of khat and one's life situation is highlighted by interviews and participant observation conducted with young male entrepreneurs. Khat in itself does not produce the dreams and desires that I have described above, and the experience of khat does not exist independently from the condition in which it is chewed. The act of working appears to focus one's mind in a way that prevents the dreaming experienced by unemployed chewers. Working young men in my study were generally entrepreneurs who worked as street side bicycle mechanics, barbers, petty vendors, and similar occupations. These entrepreneurs chew khat almost daily. Despite their habitual khat consumption, they were very critical of those who simply sit and chew without working. Like non-chewers, they referred to unemployed chewers as dreamers. Some entrepreneurs claimed that they never chew khat when they are not working because they want to avoid dreaming and speculating about their future. A group of young men who sold and repaired watches along side the street explained that,

We only chew khat at work. We never chew if we are simply spending time at home. The reason is that khat opens your mind and creates all sorts of new ideas and dreams. A person will imagine himself as the owner of a car or a hotel. Sitting at home, chewing gives far too much space for dreaming of a spectacular future, and no way to act on it. This is not good. Work focuses your mind and energy. If you chew during work the mind is only partially opened.

The contrast between working and unemployed chewers is similar to Ezekiel Gebissa's discussion of rural and urban chewers (2004, pp. 7-10). Working young men are very similar to rural chewers in that their periods of chewing are shorter and always followed by intensive work. The dislike for "dreaming" among working young men in my study is remarkably similar to a quote from an Oromo farmer in the 1960's who states that if khat "chewing is not followed by hard labor, it serves as an irritant rather than a stimulant" (Gebissa, 2004, p. 9). Working young men's experience of *mirqana* is significantly different than that of unemployed chewers. For those who work and chew, *mirqana* provides a powerful energy that is useful for performing repetitive labor and problem solving.

One bicycle mechanic described a situation in which he needed to straighten the forks on a bicycle. Especially with extremely limited tools, this is a very difficult task. As he sat chewing and contemplating the problem the young man developed a solution. He placed the forks directly in the sun and let the intense afternoon heat soften the metal. He then used a large flat rock to pound the forks back into shape. He proudly told me that for work that would have cost 20 birr and required complicated equipment at a professional garage, he charged 3 birr and did the work using only the sun and a rock as tools. Khat was perceived as providing the creative energy necessary to solve this problem.

In the same manner that khat may reproduce the condition of being jobless, it is perceived as providing the energy and creativity that allows a young worker with inadequate tools effectively to compete with larger establishments (for example, a garage). Khat use is also shaped by the condition of being a young entrepreneur in the informal economy. The control over one's time, the lack of a need to present a respectable image, and the belief that khat enables one to work effectively all encourage the young entrepreneur to chew daily. The contrast between the experience of khat for unemployed and working young men demonstrates that khat consumption is inextricable from context. Although khat does consistently produce a similar biological response among consumers, this response is meaningless outside of the particular economic and social situation of the consumer.

IMAGINING YOUTH IDENTITIES THROUGH KHAT

I have argued that young men use khat to imagine possible futures and to think through problems they face in both work and unemployment. Among both adults and young people, a counter discourse concerning khat consumption is present. In arguing that khat is a social problem, this discourse serves to construct the meaning of youth as a social category in Ethiopia.[8] In an article in the Addis Ababa based English paper, *Fortune*, on the closure of khat houses in Addis Ababa, Dr. Nigussie, a professor at Haramaya University, states, "No question about it that [khat] is a stimulant, it kills work enthusiasm, wastes time, deflects the working force from productivity, shortens one's life span, and thus it hurts the economy as a result" (Negatu, 2004, p. 3). Dr. Nigussie's claim represents the widespread belief that khat is the force behind unemployment and the idleness of youth. It is khat that has created an unprecedented population of unemployed young people and the resulting social problems.

Many young people also express this discourse. The following quote comes from an unemployed young man who did occasionally chew khat with his friends but who was critical of daily chewers.

The good thing about khat is that it helps you pass the time. If you are just wandering around alone people might talk about you and it will be obvious to everyone that you do not have work. With khat, you can be with your friends and stay in one place. This is a good thing to do from time to time but if you chew everyday you will adapt to khat and become lazy. I know people who are addicted to khat. If they have five birr, they spend one birr on *shuro* [a spicy chickpea paste] and four on khat. Good food is necessary if you are chewing, but these khat addicts do not eat properly. They only spend their money on khat and eventually they become a *jezebah*.⁹ These people are very difficult to be around if they don't get khat. Waiting between chewing is good because you don't become dependent on khat. You can take advantage of the strength that khat may give and save it for when you need to do a lot of work. Khat stops people from doing the things in life that are necessary. Khat chewers do not grow. They will never leave their parents' home, marry, or find better work. With khat progress is impossible.

In addition to describing the dangers of khat dependence, this young man claims that khat prevents progress. The particular aspirations of young men are made impossible by khat consumption. Khat undermines the potential for youth to take an active role in their life and work towards attaining their progressive aspirations. In this discourse, youth are not creators but passive receivers of information. Youth are constructed as lacking agency. Khat acts on youth, and youth are not capable of using it for their purposes.

Discourses concerning youth and khat that construct youth as lacking agency are implicitly opposed by the symbolic role of khat in relation to identity for some young chewers. Young consumers of khat often construct identities for themselves that are a means of re-imagining the social category of youth and imbuing it with greater power to act. Young chewers explained that they consider themselves to be more knowledgeable and experienced than non-chewers. Non-chewers sometimes insult chewers by calling them "*duriye*," and chewers often refer to non-chewers as "*farrah*" (rural or backwards; the opposite of cosmopolitan). This is not because rural Ethiopians do not chew khat, but because in an urban context strongly influenced by Orthodox Christian cultural norms, khat consumption connotes a certain amount of youthful rebellion. Young chewers claimed that they are more likely to be sexually active and to drink alcoholic beverages than non-chewers. In general, young chewers see themselves as "*arif*" or "*arada*." Both of these terms are common in the slang spoken among youth in urban Ethiopia and they have similar connotations to "cool" in American English.

Chewing khat is "*arif*" because it is an act that demonstrates the power of young people to act in opposition to dominant social norms.[10] It is interesting that like Neil Carrier's (2005) study of khat (*miraa*) consumption in Kenya, khat is considered cool and rebellious despite its importance for traditional religious and cultural ceremonies in Ethiopia. Carrier argues that khat is "cool" (*poa* in Swahili) because it is generally associated with a youth culture that embraces hybrid cultural practices and because it is increasingly condemned by authority figures. A similar process appears to be related to the positive evaluation of khat consumption by a subculture of young men in urban Ethiopia. In the face of claims from academics, politicians, and their parents that khat is a social scourge, young men continue to chew it. In doing so, they are implicitly opposing the construction of youth as lacking agency. In the sense that chewing is an act of defiance, it is also a display of the agency of youth. It is a sign that, despite difficulties in finding work, obtaining financial independence, and starting a family, young men still have the power to act.

CONCLUSIONS

For young men khat is both a tool and a sign. As a tool, young men manipulate the psychoactive and social effects of khat in order to negotiate a world in which they face serious difficulties in taking on the normative social responsibilities of adults. As a sign, young men use khat to signal their power to act in opposition to social and cultural norms. Both the symbolic and functional efficacy of khat depends on its particular fit with the context in which it is consumed.

Young men in urban Ethiopia chew khat in a context of unprecedented youth unemployment that has created massive barriers to attaining one's aspirations. Khat consumption generates discussions of imaginative future possibilities. Khat fueled dreams of the future serve as a means to escape the present and to facilitate dialogue about day-to-day problems that young men encounter. In contrast, for working young men khat is not a source of dreams but a means to generate physical energy and mental creativity that are essential for success in the urban informal economy. In both cases, the particular function of khat conforms to the environment in which it is consumed.

A prominent discourse among both adults and young people denies the creative role of khat consumption. Khat is perceived as shaping youth behavior rather than as a tool that may be utilized by youth. In this discourse youth are constructed as passive and as lacking agency. Many young male chewers construct identities that are oppositional to this perspective. For these young men khat is a symbol of rebellious cosmopolitanism. Young men construct identities in which khat consumption signifies a break with social norms and is evaluated positively. Implicit in this move is that youth do possess agency, if only to act in an oppositional manner. Khat consumption provides a simple and direct way in which young men may assert their presence as a unified group that is capable of acting on the world.

Finally, it is important to note that the symbolic and functional properties of khat represent continuities and disconnections with the manner that it has been traditionally consumed in Ethiopia. Young men's use of khat to explore the future resembles the use of khat by Muslim elders to negotiate marriages and solve conflicts. In both cases the properties of khat as a stimulant and social facilitator are utilized to solve relevant problems. The symbolic nature of khat, however, has changed. In the past khat was an important symbol of Muslim identity, but this is not the case for young men in turn of the twenty first century urban Ethiopia. Rather, to chew khat is to express an allegiance with urban youth culture and express a certain degree of rebellion. Khat continues to be consumed in order to solve problems and signal identity, but the nature of those problems and identities has changed.

NOTES

1. The research for this chapter was funded by a Fulbright-Hays Fellowship and grants from the Emory University Institute for African Studies, the Emory University Internationalization Fund, and the Emory University Department of Anthropology. Thanks to the African and African American Studies Program at Washington University in St. Louis for support while preparing the final version of this chapter. I am grateful to the participants of the Workshop on Khat and the Ethiopian Reality at Addis Ababa University where an early version of this paper was presented. A special thanks goes to Ezekiel Gebissa for his insightful comments and meticulous editing.

2. The importance of khat consumption for ethnic and religious identity in Ethiopia varies depending on region. In some places khat may act as a significant marker of one's identity. For in-depth analysis of the relationship between ethnic, religious, and national identities in Jimma, see Mains 2004.

3. Italics denote words that were spoken in English.

4. This appears to be partially related to notions of public space in Jimma. To wander around ("*mezor*"), especially in the afternoon, is associated with being a "*duriye*" (delinquent). To be seen in public like this frequently leads to insults. One of the advantages of khat is that it is chewed in private and it consumes a large amount of time. Although one may be unemployed, at least his idleness will not be visible if he is hidden away chewing khat. Some of the few unemployed youth who did not chew khat resented the fact that because they spent more of their time in public areas they were considered to be *duriye* (See note 7).

5. The association between khat and sex is certainly not universal. Many chewers actually claim the khat causes temporary impotence. For a detailed discussion of khat, sex, and HIV. See Abebe et al., 2004.

6. It was not simply the fact that I lived in Atlanta prior to conducting research that caused these young men to reference the city. All three of these young men had family living in the Atlanta area.

7. *Duriye* is an important term for discussing unemployed youth in Ethiopia. It is used very broadly to describe idle youth who engage in socially unacceptable behaviors like smoking cigarettes or heavy drinking as well as more serious activities like theft. It is not common for youth to refer to themselves as *duriye*, but most will at least acknowledge that others, especially adults, may call them *duriye*.

8. Debates concerning the consumption of khat also serve discursively to determine who is included in the category of youth. Through discussions of khat, "youth" is constructed as socially deviant urban men (Mains 2007).

9. A *jezebah* is someone who has been driven mad due to the consumption of khat. Young men often claimed that particular mad men who wandered the streets dressed in rags were previously healthy and had been done in by excessive khat consumption.

10. It is important to note that not all young chewers valorized khat consumption. Many felt a profound sense of guilt about their

Chapter 4

Keeping Tradition and Killing Time: The Use and Misuse of Khat in Ethiopia[1]

Ezekiel Gebissa

For nearly a millennium there has been socially acceptable ways of khat use and stern opposition to any kind of its use in Ethiopia. This tension is evident in the first known reference to khat use in which Sultan Sabradin of Ifat who was at war with the Emperor Amde Seyon is quoted as saying: "I will take up my residence at Mar'ade, the capital of his [Amde Seyon's] kingdom and I will plant [khat] there because the Muslims love the plant" (Huntingford, 1965, pp. 55–56). The sultan's statement refers to a plan to "appropriate the land of the Christians, transform their agriculture and impose on them the Muslim way of life—all of which are represented in here by [khat]" (Weir, 1985, p. 71). He may well have succeeded in introducing to central Ethiopia a plant whose production has had tremendous impact on Ethiopia's agriculture and whose consumption many view as a "Muslim" habit.

By the mid nineteenth century, according to contemporary accounts, people in central Ethiopia chewed the fresh leaves or used them as an astringent medicine, boiled with milk or water, and drunk as a beverage. The leaves were also plucked and dried in the sun, most probably for uses other than chewing or for export to distant regions or for other non-recreational purposes (Harris, 1844, pp. 334-335). In southern Ethiopia, Charles Beke, the English traveler, describes khat as "a favorite Arab intoxicant used as some sort of tea," and suggests that an infusion was made out of khat and consumed as beverage (Beke, 1843, p. 263).

While khat use was well-established in parts of central and southern Ethiopia by the mid nineteenth century, the main area of the khat chew culture remained the region around the city of Harer. The political elite, religious devotees, and well-off urbanites of the city chewed khat on a variety of occasions and on a daily basis in the city of Harer. Richard Burton, for instance, observed a khat chew party in Harer at the residence of the Treasurer of the Emirate of Harer in which several dignitaries, including the chief of government, participated. At this chew session, which took place between 9 a.m. and noon, fresh khat leaves were pounded in a wooden mortar and the paste was distributed among the participants who rolled them into chewable bales. The method of choice, however, was to pluck off the more tender leaves and then chew. Also recorded in Burton's account was the Muslim *ulema*'s notion that khat was "food for the pious" and chewers' explanation that it had special properties of "enlivening the imagination, clearing the ideas, cheering the heart, diminishing sleep, and taking the place of food" (Burton, 1987, p. 31).

Three decades after Burton, Mohammed Moktar, an officer in the Egyptian army, reported that khat chewing had become a commonplace practice among pious Muslims of the city and its environs. His account shows that khat chewing had developed to a ritual with an elaborate etiquette and with close connection between chewing and worshiping Allah (Moktar, 1876, pp. 369-372). In addition to confirming that, by the mid nineteenth century, khat use had become a cultural practice in the Muslim-inhabited regions of eastern-central Ethiopia, Moktar's description identifies at least five important factors that distinguish khat use in the nineteenth century from the contemporary practices. First, the method of khat use was not limited to chewing. The leaves were boiled, ground, and used as a beverage or for medicinal purposes. Second, the political and social elite of Harer used khat regularly in chew sessions in ceremonies closely associated with religious observances. Third, the chew took place in the morning or at night during worship, unlike the contemporary *bartcha* or the afternoon chew session. Fourth, khat was consumed in groups and the chew sessions followed prescribed courtesies, etiquette, and good manners, all of which served to socially regulate its use. Fifth, khat chewing was not mixed with alcohol. These features contrast with the chew session observable today in Ethiopia's urban areas, where khat is used primarily for pleasure and mixed with other substances.

In recent decades, khat chewing has become a ubiquitous pastime in Ethiopia's urban centers. The rate of its expansion and its universal appeal to all social groups has alarmed many people who view the practice as a sign of social decadence. This chapter aims to shed light on the debate over khat by investigating the extent of khat consumption and examining the discourse on the legal status of khat in Ethiopia. It begins by documenting the expansion of khat, shifts to a critical appraisal of arguments for the prohibition of khat use, and outlines khat's socio cultural significance based largely on data obtained from Harerge highlands, the traditional area of khat consumption and production. The purpose of focusing on the sociocultural functions of chewing is to explore why people choose to chew the leaves, for, ultimately, it is behavioral change, rather than legislation or the medicalization of what is essentially an issue of personal choice, that will determine the formal status of khat in Ethiopia. The chapter approaches the subject from the vantage point of the people whose livelihood is connected with the industry.

EXPANSION OF THE CHEW CULTURE

Until the Italian occupation of Ethiopia (1936-1941), the consumption of khat for pleasure remained a practice confined to the relatively affluent, mainly urban, predominantly Muslim population in eastern Ethiopia. In the decades that followed, khat consumption spread throughout the population nationally and became the commonplace pastime activity of nearly all social classes, cultural categories, and religious affiliations. By the late 1990s, the proportion of khat chewing had reached 70 percent for men and 35 percent for women in a predominantly Muslim community in south central Ethiopia (Alem, Kebede & Kullgren, 1999, p. 87). In a community of mixed religions in southwestern Ethiopia, current khat chewing was 40.4 percent of men, 18.2 percent of women (Belew, Kebede, Kassaye & Enquoselassie., 2000, p. 15). According to one representative national assessment of 16,606 adolescents and young adults, frequent khat use in varying degrees is evident in all regions of the country (Kebede et al., 2005). In Addis Ababa alone, the amount of khat consumed in the city each day increased by 79 percent, rising from 319 tons in 1980 to 571 tons in 2000. Surveys show similar increases around the country (Anderson et al., 2007, pp. 66-72).

In contemporary Harerge, only a tiny minority of the population, owing to personal and medical reasons, refrains from chewing khat. The sight of civil servants, day laborers, farmers, students, truckers and cabbies, shopkeepers, couriers with bulging cheeks filled with khat, is a common phenomenon in major cities in the region. Such impressions are supported by research. For instance, a 1991 survey of households in the Habro district of Harerge found that all the male household heads and 81 percent of the population between the ages of 15 and 30 years consumed khat regularly. In nearby Gelemso secondary school, the survey showed that 88 percent of the female students and 96 percent of the male students use khat on a regular basis (Feyissa & Aune, 2003, p. 188).

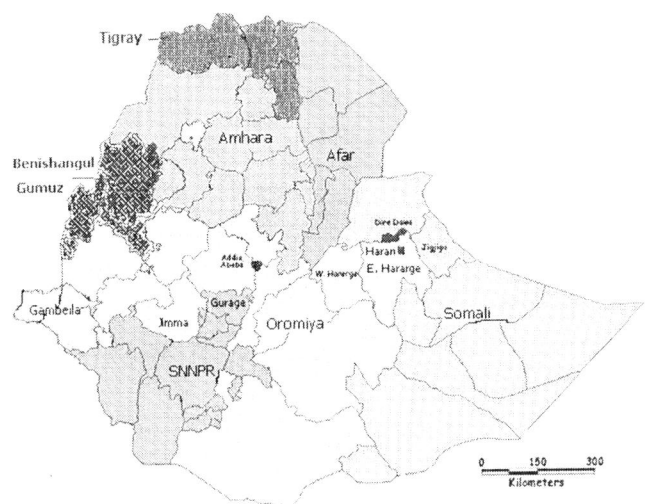

FIGURE 4.1: ADMINISTRATIVE REGIONS

Despite the negative attitudes of nonchewers, the ambivalent policy of the Ethiopian government, and the dire warnings of the medical community, khat consumption has spread from Harerge to all parts of Ethiopia's Oromia Regional State in the last three decades. A survey of secondary school students in Agaro, Western Oromia, for instance, revealed a current khat chewing prevalence of 65 percent and that two thirds of the most frequent users were students between the ages of 15 and 22 (Adugna, Jira & Molla, 1994, p. 122). A cross-sectional survey conducted in 2003 at Jimma University, located in a larger town about 30 miles to the west of Agaro, showed a lifetime and current prevalence of 46 and 39 percent, respectively. The current prevalence at the university was similar to that of Jimma town (30.8 percent) and comparable to that of Adamitulu (31.7 percent), located about 100 miles to the northeast (Gelaw & Haile-Amlak, 2004, p. 182).

Beyond the Oromia region, khat consumption has expanded to the Southern Nations, Nationalities and Peoples Region (SNNPR), particularly in the main khat farming areas of Sidama and Gurage Zones. In Butajiraa, a rural community southwest of Addis Ababa, 56 percent of respondents reported lifetime chewing of khat with 90 percent reporting daily chewing (Alem, Kebede & Kullgren, 1999, p. 88). In the Amhara region in the north, the conservative bastion of Ethiopia's Christian heartland where khat had been despised until recently, khat consumption has become a ubiquitous pastime in the city of Bahir Dar and surrounding districts. A study on the incidence of khat chewing among college students in the Amhara region revealed a prevalence of 27 percent lifetime chewing with 46 percent reporting the onset of the habit during their senior year in secondary school (Kebede, 2002, p. 13).

These regional surveys are consistent with the findings of countrywide surveys. Data from a survey of 3,700 respondents conducted in 1996 in Addis Ababa and 24 towns across Ethiopia showed a prevalence of about 30 percent and that the use of khat has spread among all segments of the population (Arrafaine, 2004). A house-to-house survey in rural areas of Ethiopia gave a comparable figure of 32 percent (Belew et al., 2000, p. 32).

Alarmed by the rapid spread of khat cultivation and use, opponents have launched a campaign in favor of proscription of khat use. Some regional states, including Tigray, Benishangul, and Gambella, have adopted local ordinances aimed at controlling purported deleterious consequences of khat. Interest groups are being organized to lobby the government to outlaw khat before, as they say, it destroys the youth and the future of the country (Anderson et al., 2007, p. 71).

Apart from those who oppose khat because of a personal conviction that khat is offensive to their sensibilities or an affront to their faith, the struggle over khat has always been between two camps: those who believe that khat's economic benefits outweigh the negative social consequences on the one hand, and those who maintain that khat is the cause of so much indolence, violence, and social decadence that no amount of economic benefit can justify its continued use. To bolster their position, the latter have often cited "scientific" studies which, from their perspective, have demonstrated the harmful medical effects of khat. In the majority of cases, the oft-cited studies are based on surveys of consumers who tend to chew outside the traditional setting and are more likely to misuse khat. Focused on proving the allegedly deviant behavior of chewers, the earlier studies have indeed provided fodder for the opponents.[2]

HISTORICAL PERSPECTIVES

In Ethiopia, khat gained notoriety among antidrug elements during the last century. Many of the critics were Christian rulers and settlers from central and northern Ethiopia who established their suzerainty over the local Muslim Oromo, Somali, and Hareri peoples in the late 19th century. For the Christian rulers of the Oromo inhabitants of Harerge, khat consumption was a mark of apostasy. In 1916, *Lij* Iyasu, Emperor Menelik's designated successor, was deposed, among other reasons, for indulging in khat chewing with Muslims and apparently converting to Islam during one of his visits to Harer in the early 1910s. He reportedly made unannounced visits to the French Somaliland and completely emptied the funds of the Ethiopian legation there in order to indulge in his habits (Ali, 1985, p. 12). Khat was considered so evil that *Fitaurari* Teklehawariat, a supposedly enlightened governor general of Chercher, Western Harerge, ordered the Oromo to uproot their khat plants and replace them with coffee trees. He placed a high tax on khat transactions to restrain market growth. He justified his actions by pointing to the apparent debilitating effect of khat on the human body and the resultant reduced productivity, diminished reproductive capacity, and increased incidence of mental illness. His efforts at eradication failed as local opposition proved too strong for him to be able to implement his plan. In the 1980s, another governor of Harerge, Colonel Zeleke Beyene, citing the negative social consequences of chewing, launched a campaign to eradicate khat. As was the case with his predecessor, the colonel was forced to abandon his campaign (Mulatu & Kassa, 2001, p. 104).

Though khat chewing was associated with Islam, Muslims did not always hold a uniform view regarding the legitimacy of chewing. In the early twentieth century, some Islamic *ulema* in Harer expressed opposition to khat use on religious grounds. The *imam* of a mosque in Dire Dawa emerged as a vociferous opponent of khat chewing, arguing that it was a mind-altering substance that the Prophet had expressly condemned along with alcohol. Despite his vigorous public campaign against chewing, khat consumption continued to increase. Perhaps most Muslims agreed with other leaders who argued that khat was a *katumomia* or that which God blessed and gave to men through his favor. Those who participated in religious ceremonies held at Muslim shrines in the province spent long hours chewing khat while reciting verses from the Quran and praying to Allah during Ramadan and at the *Arafa* and *Mawlid* celebrations (Brooke, 1960, p. 53).

Present-day opponents do not contend that khat has absolutely no benefit for society. They recognize that, from a macroeconomic perspective, the increased money circulating in the khat sector does promote growth and provide employment. The government collects large sums of foreign exchange earnings and revenues from taxes on domestic transactions and uses khat revenues for public expenditure (see Chapter 5). The objection against khat use is based on the notion that khat is an addiction producing drug with adverse health effects and deleterious socioeconomic consequences for the chewer.

ADVERSE HEALTH EFFECTS

Even though the claim that khat consumption entails adverse health consequences is often exaggerated (Kennedy, 1987), there is broad agreement that khat chewing can lead to increased body temperature, pulse rate, blood pressure, and respiratory rate (Halbach, 1979; Hassan et al., 2000; Kalix, 1987). The presence of tannins in khat leaves is also identified as the probable cause of gastrointestinal tract problems, digestive disorders, and the duodenal ulcer occasionally reported by users (Dhaifalah and Šantavý, 2004; Pantelis, Hindler & Taylor, 1989; Raja'a et al., 2001). Khat has been associated with impotence and lower semen parameters and considered harmful to reproductive health (El-Shoura et al., 1995). Researchers have also found that khat can lead to psychological problems with symptoms such as sleeplessness, nervousness, and nightmares. Habitual chewing can result in functional mood disorder, depression, and other symptoms requiring vigorous treatment. In general, khat might precipitate psychiatric disorders in vulnerable subjects and it can exacerbate symptoms in patients with preexisting conditions (Yousef, Huq & Lambert, 1995).

While these adverse health effects of khat have been described in numerous studies, researchers often cautioned that their findings are preliminary indications that should not be projected or construed as firm conclusions. Though earlier studies were apt to warn against negative health consequences of khat chewing, they lacked data on the prevalence of adverse effects and the results of controlled studies that compared chewers and nonchewers. Later researchers made up for these deficiencies but were surprised to find that the majority of the health consequences may not be directly attributed to khat (Kennedy, Teague, Rokaw & Cooney, 1984; Kennedy, pp. 193-232). A careful reading of the latest comprehensive review of the literature on the medical and pharmacological effects of khat confirms that the findings concerning the impact of khat are inconclusive and contradictory. While some studies have shown khat to have negative effects on fertility, others have refuted such an association. For instance, a study conducted in Yemen, a country where nearly all the adult male population chews khat, shows the country "has the highest fertility and annual population growth rate in the Eastern Mediterranean region" (Al-Hebshi & Skaug, 2005, p. 303). The same is true for the Harerge Highlands where in some places population density reaches 420 persons per square kilometer (Wakjira, 1989). Likewise, while many maintain that khat causes psychological disturbances, others see no direct association between chewing and any form of psychosis. Further, a study conducted among Yemeni immigrants in Israel found a higher rate of periodontal disease, but a study carried out in Kenya showed no significant difference in the periodontal health of chewers and nonchewers (Al-Hebshi & Skaug, 2005, p. 304). Given the inconsistency in the literature, the best conclusion that can be drawn is one that Luqman and Danowski, two of the first scholars to identify a list of khat-induced health problems, made three decades ago. They conclude that, "in the absence of more definitive information, one can only cite clinical observations and continue with hypotheses concerning clinical disorders in the users of khat." (Luqman & Danowski, 1976, p. 248)

Yet, opponents of khat use in Ethiopia seem to have no qualms disregarding the nuances and lacuna in scientific knowledge and casting khat in ominous terms. The list of ailments that opponents ascribe to khat gives one the impression that no part of the human body, from the hair to the toes, escapes the alleged dangerous effects of the "evil" leaves of the "cursed" tree. Khat seems to affect the nervous, digestive, respiratory, reproductive, cardiovascular systems, and other functions of the human body (Woldemichael, 2003). While it is evident that regular use of khat entails some side effects, it is worth noting that opponents stretch the evidence to hold khat culpable for the abuse of other substances. Notice what Worku Woldemichael (2003), an outspoken proponent of a complete ban, has to say:

> In addition to the effects of khat per se, some substances consumed together with it are also known to produce significant adverse effects by themselves. One such substance is tobacco taken as a cigarette smoke for brain stimulation. The consumption of tobacco is associated with a number of serious adverse effects including cardiovascular and respiratory disorders, lung cancers, anorexia and addiction. (Woldemichael, 2003, para 9)

There is no debate about the adverse health effects of cigarette smoking or that some khat users smoke cigarettes. However, despite the popular belief that cigarettes and khat are inextricably linked, studies show that the majority of lifetime khat users are not cigarette smokers (Kebede, 2002, p. 9). Those who chew as a matter of tradition, for worship, or to put the high of the *mirqana* (the high) to work do not necessarily smoke during the chew session. For rural chewers, khat is chewed for the energy it gives them for the arduous task of farm labor and are always dumbfounded when told urban chewers use alcohol to expedite the wearing off of the *mirqana*. As we will discuss in detail below, this is an important distinction between khat users and "abusers," but opponents often tend to blur them and do not hesitate to conjure up a more dangerous association:

> The concomitant use of alcohol to counteract the stimulant and insomniac effects of khat raises the risk of alcohol abuse. . . . Recently, it has been observed that people with alcohol-use disorders are more likely than the general population to contract HIV. It is therefore possible that the use of khat can promote this process through alcohol consumption, among other possibilities. It is not also difficult to hypothesize that in persons already infected, the combination of khat use and HIV can be associated with increased medical and psychiatric complications. (Woldemichael, 2004)

Even though a scholarly study on possible khat-HIV association found HIV infection rate to be higher among chewers (59 percent) than nonchewers (41 percent), the finding of a single study cannot be a sufficient basis for the kind of extrapolation the khat opponents have made.[3] The study (Abebe et al., 2005) is premised on the notion that the use of khat in combination with alcohol removes the socially imposed discreetness people exercise under normal circumstances. The absence of restraints could result in risky behavior and vulnerability to contracting the virus. However, the study does not establish that khat chewing directly leads to increased alcohol consumption and heightened sexual drive. A survey of more than 20,000 in-school and out-of-school students showed "a substantial proportion of out of-school youths engage in risky sexual behaviours and that the use of khat, alcohol and other substances is significantly and independently associated with risky sexual behaviour among these young people" (Kebede et al., 2005, p. 7). Here again, the study did not have a control group to determine whether the sexual behavior of nonchewers who are out-of-school students is different from the studied group. As indicated earlier, traditional consumers do not mix alcohol and khat and a significant portion of contemporary urban users[4] rarely use alcohol with khat (Green, 1999, p. 40). The khat-HIV connection is tenuous at best and the projection of HIV risks onto khat users by opponents outlandish. By trying to link all kinds of health problems with khat consumption, the opponents' purpose is clear. If khat is shown to be the culprit, it becomes easy to make a more compelling case by relating poor health to the larger issue of negative socioeconomic consequences, thus rallying an indifferent public behind the cause of outlawing khat.

DELETERIOUS SOCIOECONOMIC CONSEQUENCES

Observers suspect that the increasing ubiquity of the chew culture is a consequence of rampant poverty requiring government intervention. The basis of this assertion is that khat chewing is a habit taken up by the urban unemployed as a means to escape the depressive effects of joblessness (see Chapter 3). Thus far, not a single study conducted in Ethiopia has established a direct correlation between poverty and the onset of khat chewing. The main reason mentioned by khat chewers as a reason for commencing khat chewing is increased performance, followed by pastime entertainment, conviviality, and socializing. Even studies of the sociodemographic correlates between khat and income show that the majority of the people taking up the chewing habit are economically better off than others (Gelaw & Haile-Amlak, 2004, p. 181). The argument that khat is symptomatic of a condition of poverty is not borne out by empirical data.

In the case of khat's alleged negative socioeconomic consequence, the popular argument depicts chewers as lethargic individuals who spend most of their days masticating on the leaves. The implication of such an assessment is that their jobs often suffer. The sight of *bartcha*, the afternoon chew session, inevitably impresses upon observers, including some scholars, that khat is a cause for tardiness to work, absenteeism, and declining productivity (Gelaw & Haile-Amlak, 2004, p. 182; Kassaye et al., 1999, p. 105). To be sure, there has been little or no scientific study of the impact of khat-induced tardiness and absenteeism on productivity or the number of lost man-hours in agricultural labor in Ethiopia. The few published surveys of users that exist suggest that khat augments energy levels, enhances imagination and creativity, increases alertness and confidence, and facilitates communication.

If it is true that khat chewing causes productivity to fall, the impact should be felt in the agricultural sector. Accordingly, some researchers have posited that farmers who grow khat waste their labor power on growing a harmful drug instead of on producing essential food crops. In this regard, there is an obvious disconnect between what farmers believe and what khat opponents tend to project. Farmers have consistently reported to researchers and casual observers alike that "the stimulant leaf gives them energy and strength to accomplish their strenuous agricultural activities, which would otherwise be impossible" (Lemessa, 2002, p. 4). This has led some researchers to conclude that khat farmers are more efficient than cereal farmers (Mulatu & Kassa, 2001). The existing evidence does not tell us whether khat chewing increases work efficiency, but the amount of labor employed in khat production is miniscule at the macrolevel, contrary to the widespread perception. Data obtained from the fifth round of the Ethiopian Rural Household Survey (ERHS) demonstrates that 6,983 households invested labor amounting to 341,650.60 man-hours in the production of various crops, including khat, out of which only 4891.51 man-hours (or 1.43 percent) went into the production of khat. In contrast, 31,542 man-hours (or 9.23 percent) were employed in coffee production (Arrafaine, 2004).

It is also argued that individuals divert their income to khat at the expense of their family needs. The resultant neglect is often cited as a causal factor in family discord and divorce. The argument here is that habitual users spend as much as one half of their income on khat and hence cut severely into the family budget, taking "food from the mouths of children and cloth off the women's backs" (Green, 1999, p. 41). While there appears to be no quantitative study of the impact of income diversion on the family and the incidence of khat-induced divorces, there is widespread perception in Ethiopia that khat chewing is a major source of family breakup and child destitution. Aware of the evocative power of moral reasoning, the opponents of khat were quick to raise the specter of greater moral inequity:

> It becomes unhealthy for the society if a few individuals benefit at the expense of many other members. If the [khat] farmers and traders also happen to be consumers, inevitably, in the long-run they will also be more directly affected by the drug, although in the short-run they may be in a much better shape than the mere consumers due to the compensation they obtain from the profit. (Woldemichael, 2004, p. 9)

The thrust of the moral imperative is obvious. Producers and marketers are benefiting at the expense of consumers and the transfer of wealth from one social group to another is inherently immoral.

Critics have also argued that khat is a "gateway drug" to other highly addictive drugs such as cocaine and opiates. In this respect, even authors of scientific studies, wittingly or unwittingly, take part in the dissemination of popular opinions about khat that their own data contradict. For instance, Kassaye et al. (1999) states:

> It is obvious that drug use has negative consequences on the economic development of a country as the health, time and money of the most productive section of human resource [youngsters] are affected by the habit of indulging in drugs... if a policy is not formulated that regulates the use of illicit drugs, including khat, the use of hard drugs among youngster [sic] is inevitable in the near future (p. 105).

Ironically, the same study that warns against the dangers of khat shows that chewers do not progressively advance to drugs that are more potent. A survey of 36 high school students in Ethiopia's capital on which the study is based showed that "none of the students reported the use of hard drugs, such as cocaine or heroin ... [and] 23 (64 percent) admitted that the first 'drug' they took was alcohol" (Kassaye et al., 1999, p. 105). Hence, the available data do not sustain the notion that khat users inevitably take up other drugs of abuse. In fact, khat use seems to serve as a protective mechanism for young chewers who live in areas where the opportunity to advance to more potent drugs is high. A panel of experts in the United Kingdom found that "Khat users do not use other drugs of abuse. It is suggested that this consequence is facilitated by khat's legal status" (ACMD, 2005).

In general, critics condemn scholars who present khat in a positive light as purveyors of ethnic conflicts. Because the majority of beneficiaries of the khat industry in Ethiopia are Oromo, studies that emphasize the economic benefits of khat are invariably characterized as parochial, ethnocentric, and unpatriotic. Woldemichael (2004) states bluntly:

> Arguments should not also be based on ethnic grounds narrowly as Ethiopia is a multi-ethnic country, which every citizen, whenever possible, is expected to offer "patriotic" contributions in order to assure its survival as an intact sovereign nation. The suggestion that the use of [khat] should not be

> regulated by central authorities because of its local cultural significance is not also an acceptable idea in view of the harm the drug inflicts on the Ethiopian society. If a cultural practice is found to be harmful, it is necessary that it should somehow change. (p. 10)

Khat opponents see an even more sinister motive in the government's reluctance to institute regulatory regimes to curtail the spread of khat consumption. The opposition to khat has joined the cause of the political opposition. In the view of these groups,

> ... the TPLF-led regime in Ethiopia is involved in the [khat] business for a couple of other reasons. By allowing the widespread use of the drug in the country, it seems to strive hard to destabilize and weaken the society. This can enhance its ability to control the population and remain in power unchallenged (Zeleke, 2004, p. 4).

As is evident in the quotes cited above, the case against khat is hardly evidence based in terms of any physical, psychological, or socioeconomic harm that chewing allegedly causes. Arguing that khat has deleterious medical, political, and moral consequences, khat's opponents are determined to coerce the government into legislating it out of existence. After a plethora of the negative consequences of chewing, their conclusion is predictable.

> For such an issue, the responsibility lies entirely in the hands of the Ethiopian government, and finally it is the government that has to make the tough choice of banning the drug in the country before it causes further damage to the already suffering society.... Ethiopia should be a [khat]-free zone (Woldemichael, 2004, p. 11).

The problem is that existing research does not distinguish between traditional khat users and contemporary khat abusers (who chew to 'kill time' and cap the chew with alcohol). The prohibitionists, following the researchers, lump all chewers together as if khat affects every individual, chewing anywhere, in the same way. Norman Zinberg, after years of studying the use of opiates by American soldiers in Vietnam and back in the United States, concluded that drug effects differ greatly depending on variations in the individual ('set') and situational ('setting') factors in which the drugs are consumed. Patterns, settings, rituals, and the significance attached to one's drug use mitigate a drug's pharmacological effects (Zinberg, 1984, pp. 172-191). These principles also apply to khat chewing.

THE USE AND MISUSE OF KHAT

In the modern controversy over khat, efforts to discourage consumption, let al.one eradicate it, have always ended in failure (Griffith, 1998; Odenwald, 2007). In the case of Ethiopia, one of the reasons for the failure to curb khat expansion is the lack of understanding of the complexity of the rationale for the development of the chew culture in the first place. Until very recently, unaware of the cultural and social significance, researchers have tended to depict khat consumption as deleterious to human health and economic wellbeing. An important shortcoming of the research findings is the propensity to ignore the importance of diversity in chew practices and the likelihood that the variations could yield different results regarding the impact of khat on the human body.

In Ethiopia, specifically in Harerge, there are at least two kinds of chewers: the pleasure-seekers (predominately urban) and the purpose-oriented chewers (mostly rural, but not limited to). In the former setting, khat chewing is often followed by an alcohol drinking binge to counteract the stimulating effect of khat. The chewers also tend to refrain from eating. This pattern of khat use may appropriately be described as misuse. Rural chewers, who constitute the majority of chewers and are devout Muslim, use the *mirqana* (the high) to work, meditate, and discharge communal obligations. In their view, mixing the sacred khat with the defiling alcohol is a direct affront to their sacred tradition and an egregious violation of the chew etiquette. In the case of the pleasure-seekers, chewing takes place in a setting where the cultural strictures and chew etiquette that limit the amount of khat chewed are nonexistent. In the indigenous setting, where khat is chewed to keep tradition, cultural formalities and cultural forces prevent the occurrence of misuse (cf. Kennedy, 1987, p. 193).

Traditional chewers maintain that khat is a cultural practice blessed by God and given to their ancestors. They do not understand why anyone wants to interfere in their sacred tradition. Suspicious of the motives of nonchewers, they tend to avoid answering any questions about khat, even when they come from the most dispassionate researcher. However, whenever they show willingness to answer the question why they chewed the khat leaf, they respond with practically uniform answer: khat chewing dispels and relieves hunger, thirst, fatigue, weariness, and even the desire for sleep. These responses indicate that many chewers associate khat use primarily with work situation, insisting that it is impossible to engage in mental or physical labor without the *mirqana*.

When asked to explain why they chew, urban chewers invariably suggest that chewing allows them to relax, to socialize, or as they say, to "kill time." Though this group chews generally for pleasure, it is important to note that a significant chunk of it constitutes purpose-oriented chewers. Their responses to the same question are varied but they echo the same theme as the traditional chewers. Students assert that they chew khat because it repels sleep and focuses the mind; businesspeople state that they clinch business deals during chew session; transportation workers assert that khat helps them stay alert; and shopkeepers emphasize that khat helps them bear up the boredom of long days in their stores (Gelaw & Haile-Amlak, 2004, pp. 181-182). In the traditional setting, khat is a luxury in times of celebration (birth), a symbol signifying a rite of passage (wedding), and an item of consolation during bereavement (Gashaw, 1996, pp. 99-101). For instance, in Harerge, when a couple enters into *naqata* or engagement, the parents signify their agreement by accepting or their disapproval by rejecting the khat, as is customary, sent to them from the parents of the groom. Other chewers relate that khat allows them to have peaceful and constructive interaction with friends and relatives. They insist that khat provides an important break for a convivial social intercourse. The calm environment that the chew session creates permits rational and peaceful discourse to take place on serious community or national issues. In this sense, khat serves a medium of communication and is an integral part of community life.

In the context of the traditional setting, khat has a critical role to play in forging social integration. In the Harerge region, for instance, khat functions as an index of social identity that cuts across social cleavages and integrates Oromos, Somalis, Hareris and other groups as a distinct group, *Ya Harer lij*. In the eyes of other Ethiopians, one of the marks of being from Harer has been the chew culture. Even immigrants to Harerge adopted the habit as soon as they moved to the province in an attempt to blend in their new cultural milieu. For traditional chewers in general, the medical arguments over whether khat is beneficial or harmful are of less importance than the social reasons for its use. As noted earlier in this chapter, historically, the most vocal opponents of khat consumption in Harerge have been denizens to the chew culture, such as Ethiopian settler landlords, who feared khat might turn their tenants and laborers into lethargic individuals. The contemporary prohibitionists' efforts, though presented as a medical concern, cannot escape being viewed as paternalistic interference of outsiders who imagine the chewers themselves are incapable of making the decision to chew or not to chew khat. The elimination of khat use could even be construed as an attempt to destroy indigenous cultural patterns and lifestyles in Harerge. A policy advocating the eradication of khat use thus deserves—at the very least—an evaluation of the reaction that such a policy might produce.

Khat chewers enjoy the chew party for a variety of reasons, not the least of which is personal gratification and pleasure. In urban settings, chewing represents a luxury that people indulge in with friends and relatives in much the same way people in Western cultures, regardless of class difference, enjoy alcohol, tea, coffee, tobacco, and a host of other stimulants. In pre-1974 Ethiopia, young people spent their leisure time at the YMCA or in the thrills provided by participation in spectator sports. School-organized events, such as inter-school or neighborhood soccer tournaments, were venues where the young met for social stimulation. The advent of military rule in 1974 eliminated all of these and substituted them with forced participation in revolutionary gatherings and officially-prescribed cultural performances which had no dimension of social stimulation. In the absence of these recourses, young people in Ethiopia took up khat chewing in small groups and in private settings (Gebissa, 2004, p. 13).

The adoption of the chew culture by young Ethiopians actually commenced during the years of "Campaign Through Cooperation" (1977-1979), when high school and college students were sent to rural Ethiopia ostensibly to distribute nationalized land among farmers. In fact, the Campaign was a scheme designed to remove student activism from urban centers. In rural settings where there were no parental supervision and no recourses to entertainment, the idle students took to khat and alcohol as their preferred pastime, a habit that the returning students brought back home. In the urban setting, neighborhood associations known as *kebele* remained the primary social venues for youth interaction outside the school. However, because of the fear these institutions inspired as the government's brutal instruments of control, they were never considered authentic venues of social stimulation. Since any other gathering outside the *kebeles* was suspected of a meeting of anti-revolutionary conspirators, young people met in private settings to chew khat. On college campuses, khat chewing became the trendiest social indulgence wrapped in the cloak of a study aid, which kept students up during long nights of concentrated cramming.

In the period after the fall of the military regime, khat chewing that had been confined to the private setting became a public phenomenon. The generation that adopted khat as a pastime activity had become the cohort running the country. It was not only able to afford khat but also free of the social stigma attached to khat in the past. Khat chewing houses, similar to a restaurant or a tea house, became thriving businesses catering to different classes of people. Private homes in upscale neighborhoods and other precincts in major cities now provide khat chewing rooms fully furnished with pillows, a variety of khat brands, and other accessories needed for a chew session (Nida, 2004).

As a pastime, the contemporary khat chew session is personally gratifying for the chewer and serves as a socially valuable time that fosters comity and cooperation. The act of chewing fresh khat in a group, surrounded by friends, creates an atmosphere of social harmony imbued with tons of generosity, pleasure, and friendship. The leaves are offered to guests as a sign of hospitality, to business partners as a gift fostering reciprocity, and to friends to establish goodwill. Apart from the serious work that gets done while khat is being chewed, the chew session provides people a time for socialization, conviviality, and tranquility. The argument that chewing is a wasted time is a broad-brush assertion based on the behavior of very few chewers. Khat can be a luxury item for rich and poor alike, particularly when the chew takes place in the traditional setting wherein the amount ingested is prescribed and its use culturally sanctioned and supervised.

There is no question that the khat constituency is large and any law that restricts the use and exchange of khat is likely to face fierce resistance. Khat is not only integrated in the life of both the traditional and urban chewers but also serves a vital role of active social integration. The prohibitionists realize the case against khat can marshal enough support only if the khat problem is defined as a medical epidemic. Apparently the proponents of prohibition assume that biomedical science trumps economics and political arguments and can force parliamentarians to outlaw khat. In this, they have adopted the tactics of the proponents of social control and those of khat opponents in other countries.

MEDICALIZATION OF A SOCIAL ISSUE

In debates about the legal status of mind-altering substances, whether it is alcohol or marijuana, prohibitionists have invariably resorted to medical arguments to bolster their claims about the hazardous health and social consequences of using them. The question of why one decides to use a psychoactive substance is essentially a matter of choice. It becomes a social problem when the behavior induced by the substance affects another person. There have been rare, if any, connections made between khat chewing and criminality and violence. Even in those rare cases the evidence is extremely tenuous at best (Alem & Shibre, 1997). Apart from media descriptions, fed by khat opponents, there has never been a cause-and-effect link between khat and criminal behavior (Armstrong, 2008, pp. 635-36). This was confirmed by a study conducted in Australia (Fitzgerald, 2009), a panel of experts in the Netherlands (Pennings, Opperhuizen & van Amersterdam, 2008) and the United Kingdom (ACMD, 2005) which found no link between khat and organized or acquisitive crimes. In fact, the UK panel concluded that the violent offender rate was lower among khat users than the general population. More relevant to the Ethiopian case, where khat opponents and the general public tend to attribute the high accident rate of Isuzu trucks, indeed the transport vehicles that make retail commerce possible, to khat chewing, the UK panel has determined that "There is no correlation between khat use and impaired driving" (ACMD, p. 19).

The moral arguments of equity and the alleged negative impact of chewing on the family did not seem to gain traction for the Ethiopian prohibitionists. They have thus resorted to the established means of medicalizing what is essentially a social (or moral) issue since prohibition would mean interfering in matters of personal choice. This tactic has a track record as an effective method for achieving social control. As Irving Kenneth Zola observed some time ago,

> Medicine is ... a major institution of social control, nudging aside, if not incorporating, the more traditional institutions of religion and law. It is becoming the new repository of truth, the place where absolute and often final judgments are made by supposedly morally neutral and objective experts. And these judgments are made, not in the name of virtue or legitimacy, but in the name of health. Moreover, this is not occurring through the political power physicians hold or can influence, but is largely an insidious and often undramatic phenomenon accomplished by 'medicalizing' much of daily living, by making medicine and the labels 'healthy' and 'ill' relevant to an ever increasing part of human existence (Zola, 1972, p. 487).

A glance at the history of the opposition to khat shows that moves to restrict the use of the leaves were driven by concerns over the alleged social effects of consumption rather than by medical evidence of harm. In Kenya, for instance, British colonial officials defined social conditions as medical to legitimize their decision to ban khat in an attempt to control the behavior of their subjects. The decision that restricted khat trade was based on claims made in a series of articles that appeared in the *East African Medical Journal* in the 1940s, even though the articles were based more on sensational anecdotes rather than empirical evidence (Anonymous, 1945b). For instance, one article recounted a tale of a khat trader who walked on a path known to be frequented by lions late at night and was devoured by the beasts until "a few fragments of shinbone" were left, ostensibly demonstrating that khat produced an "irresponsible fearlessness" (Anonymous, 1945a, p. 2). Another study presented a list of claims about the negative effects of consumption even as it admitted that "not a great deal is known about the chemical and pharmacological properties of the active principles of *Catha edulis*" (Bally, 1945, p. 3). Other articles claimed that khat caused insanity (Carothers, 1945) and poisoning (Heisch, 1945) based on a couple of cases without bothering to establish a link between chewing and the coinciding ill-effects.

Medical science was once again invoked to achieve an international consensus to ban khat at a 1983 conference in Madagascar sponsored by the International Council on Alcohol and Addictions. Most participants condemned khat as a harmful substance, but Abraham Krikorian pointed out that scientific evidence was called upon to justify preconceived ideas designed to precipitate governments into action. When dealing with the use of drugs, Krikorian noted:

> .... one becomes so obsessed with the chemical and medical details that it is forgotten that one is dealing with an essentially ethical question and one effectively cloaks this in scientific garb. One thereby avoids acknowledging, as many philosophers have warned, that a given question is more open than one might care to admit — for any position wearing the mantle of 'scientific objectivity' becomes difficult to oppose. ... The relative weight given to physiological considerations in questions such as those dealing with drug control will always be ethical, not medical or otherwise scientific issue. ... permitting the aura of science to be smuggled in for support [of one's ethical position] — does a disservice to honest public discourse. ...decisions regarding control of drugs by the state will not be rendered any easier and certainly no more just by sacrificing their essentially ethical nature on the altar of medicine or science (Krikorian, 1984, p. 164).

As Zola and Krikorian show, the question of social control in general and banning of khat in particular is in fact a political issue at the national level and a moral issue at the individual level. As we have seen in this chapter, khat prohibitionists in Ethiopia seem to follow the well-trodden path of using "biomedical evidence" to lend credibility to their essentially political or ethical agenda. Here it is important to set the medical evidence that the prohibitionists in Ethiopia present against the larger context of the movement to ban khat internationally.

Ethiopian prohibitionists insist that "the United Nations has established" that khat is unmistakably a dangerous drug (Addis Reporter, 2004; Letters from anti-khat groups to me, 2004). The fact is there is no international consensus to recognize khat as an illegal drug. International organizations have been confronted with problems associated with khat since 1935, when the League of Nations Advisory Committee on the Traffic of Dangerous Drugs discussed two technical reports on the subject of khat. In 1957, the United Nations Commission on Narcotic Drugs was asked to take up the question and make recommendations on whether khat should be internationally recognized as an illegal drug. The investigations in both cases failed to establish clear evidence of the harmful effects of khat consumption. A study commissioned by the UN produced a report in 1975 that determined cathinone to be the active ingredient that gives fresh khat its potency. Previously, cathine, which is basically ephedrine, was believed to be the main active ingredient. Cathinone is many times more powerful, similar to amphetamine in its makeup, but half as potent. In 1986 the United Nations finally added cathinone, but not the khat plant, to its list of substances that should be regulated. In fact, the United Nations Office on Drugs and Crime is quite explicit that the khat plant is a stimulant "not subject to international control" (UNODC, 1996). Seventy years after the international community first took up the issue of khat, there still is not a clear-cut answer.

Anti-khat groups in Ethiopia consistently cite that khat is a controlled substance in Western countries, specifically in the

United States. When the United States made khat Schedule I controlled substance in 1993, the decision was precipitated by the botched US military operation in Mogadishu, Somalia (Anderson and Carrier, 2007, p. 150; Beckerleg, 2008). Before that incident, officials of the Drug Enforcement Agency (DEA) dismissed khat as a substance unappealing to Western users. After 11 September 2001, DEA officials began to link khat chewing with terrorism in their public pronouncements. "It is not coffee. It is definitely not like coffee," said a DEA official in 2007,

> It is the same drug used by young kids who go out and shoot people in Africa, Iraq and Afghanistan. It is something that gives you a heightened sense of invincibility, and when you look at those effects, you could take out the word 'khat' and put in 'heroin' or 'cocaine' (Dizikes, 2009).

Setting aside media reports, unofficial pronouncements, and government issued bulletins, US circuit courts have ruled that the khat plant, as opposed to the cathinone that may be found in it, "is not a controlled substance" because "neither the U.S. Code nor the Code of Federal Regulations" lists it as such. As a legal matter, therefore, "Khat itself is not illegal. Some of the chemicals that are sometimes found in it—but not always found in it—are illegal" (Armstrong, 2008, p. 637). As Sidney L. Moore, a defense attorney in New York specializing in khat cases observes, one needs "to chew about 650 lb of khat to squeeze 1 gm of cathinone out of it" (Gardiner, 2006). That is an impossible task even for the orally dexterous chewer. The human physical condition thus limits the presumed harm even over a lifetime of chewing.

While the cathinone-khat distinction is important, the real issue is whether khat is addictive in the technical sense of the word. In 1985, the WHO Expert Committee on Drug Dependence included cathine, one of the known active ingredients in khat, in the list of "dependence producing drugs" and described its properties as similar to, but significantly less potent than, amphetamine. By the 1980s, adequate research had been done on cathinone to classify khat as causing moderate psychological dependence, but no physical dependence or definite abstinence syndrome associated with cessation of khat use has been described (Giannini et al., 1986; Kennedy, 1987; Nencini & Ahmed, 1989). Khat chewers exhibit withdrawal symptoms upon cessation of chewing which researchers interpreted as "rebound phenomenon rather than a specific abstinence syndrome" (Pantelis et al., 1989, p. 658), a kind of "cultural drug dependence" or "drug-facilitated sociability dependence" (Kennedy, 1987, p. 210; Odenwald, 2007, p. 12). By the 1990s, it seems a scientific consensus had emerged that the use of khat cannot technically be described as addictive and thus it cannot be categorized as an addiction-producing drug (WHO, 2003).

Primary among the contentions of khat prohibitionists to buttress their case for banning khat is the "food deprivation" hypothesis which makes chewers vulnerable to infectious diseases. Based on reports of some chewers an extrapolation is made that khat is an appetite suppressant which fosters malnutrition and renders chewers vulnerable to disease. Obviously khat is not something that guarantees normal continuity of life and cannot be classified as a basic necessity in the same way food is. In Harerge and plausibly in all of Ethiopia, chewers in the traditional setting never confuse khat with food or never treat it as a food substitute. In fact they chew khat before or after meals. The customs regulating khat chewing sanction food intake and the avoidance of alcoholic beverages. To the eye of the observer, khat chewers consume more rice, chicken, and milk than nonchewers (Kassa, 2000, p. 16). There is a clear disconnect between the perception of the experts and the experiences of the khat users.

Today khat wine is legally produced and sold in Ethiopia, perhaps inspired by the use of *gesho* (*rhamnus prinoides*) in the manufacture of Ethiopian mead for centuries (Merab, 1912, p. 121; Simoons, 1960, p. 115; Tadesse, 1958). This indicates the almost inexorable rise in importance of khat in Ethiopia. Having read through the scientific literature on the effects of khat and observed chewers in the traditional setting, it seems to me khat use is about as injurious to the health of its users as the use of coffee or tea. It is certainly less dangerous than cigarettes and regular use of alcohol, especially when khat is used in the traditional context where custom limits the amount one could consume in a single session (Kennedy, 1987, p. 184). So, following Susan Beckerleg (2008) and echoing the befuddlement of my own informants over the commotion about banning khat, I ask "Why the fuss About Khat?" And can a law banning khat chewing be successful in Ethiopia when history has shown that draconian measures have almost invariably failed to achieve their goals?

NOTES

1. Parts of this chapter and chapters 5 and 9 first appeared in my article, "Scourge of life or an economic lifeline: Public discourses on Khat (Catha eduis) in Ethiopia. *Substance Use and Misuse, 43, 6,* 784-802.

2. There are, of course, exceptions. The best known recent publications that promote a more positive image of khat are studies based on the situation in Yemen. See Shelagh Weir, *Qat in Yemen: Consumption and Social Change.* Kevin Rushby's *Eating the Flowers of Paradise: A Journey Through the Drug Fields of Ethiopia and Yemen* is less academic but presents a fascinating picture of the culture surrounding khat in Harerge and Yemen. The best comprehensive study on the science of khat is John G. Kennedy's *The Flower of Paradise: Institutionalized Use of the Drug Qat in North Yemen.*

3. The study also found that HIV cases were higher among the married than the single and among Christians than non-Christian subjects. This fact alone, however, did not lead the researchers to the conclusion that marriage and Christianity make individuals vulnerable to HIV.

4. According to Reginald Green's estimate, Ethiopia produces khat on the order of $400 to $500 million at wholesale prices. A third of this is exported to Djibouti and Somaliland, and the bulk consumed in the Somali Regional State and by the Somali community in Addis Ababa. A significant amount is also consumed in the Oromia Regional State. In nearly all of these areas, the consumers are Muslims who do not mix alcohol with khat.

PART II

ECONOMICS OF PRODUCTION AND TRADE

Chapter 5

Crop and Commodity: Economic Aspects of Khat Production and Trade

Ezekiel Gebissa

ಜಲ

Ethiopia has often been characterized as a monoproduct export economy that failed to achieve self-sustaining growth. In the late 1940s and early 1950s, coffee was a major cash crop dominating farm fields and domestic commerce, and the primary agricultural commodity linking the country to the world market. Inside the country, the coffee sector was an important source of income for farmers and workers who performed the selection, processing, and packaging the commodity before it could be dispatched to international markets. These economic benefits depended on high coffee prices, which in turn were determined by favorable rain patterns and absence of incidence of coffee disease. Yet there was no attempt to diversify the export economy in a systematic way (FAO, 1995). During the imperial period (1941-1974) coffee was the unrivaled "backbone of Ethiopia's economy" as long as it had frost as a friend in

Brazil. Under the Derg regime (1974-1991), it was celebrated as a "revolutionary crop, the foundation of the economy, a source of revenue, and financier of development."[1]

In the Harerge highlands from where coffee was exported mainly to the politically volatile Middle Eastern markets, the uncertainty surrounding the production of coffee led to an innovation at the farm level. In the early 1950s, khat suddenly emerged as a profitable cash crop, challenging coffee's dominance in the field and as an export commodity. Khat's profitability was linked to the proximity of the Harerge highlands to Djibouti and Somalia where it was consumed and to major highways connecting the production areas to Dire Dawa, the terminus for the Ethio-Djibouti railway. With the beginning of the Ethiopian Airlines operation in the late 1940s, air-transported khat found an important market in Aden, increasing both demand and production. As the Ethiopian government tightened regulation and taxation on coffee exports, particularly in the 1980s, following the Land Reform Act of 1975 and enactment of restrictive policies on agricultural marketing, farmers increasingly shifted to the production of khat as their main cash crop.

Parallel to this, domestic khat consumption increased dramatically. As noted in chapter 4, consumption spread not just to areas where khat was never chewed before but also to sections of society that had not been known to chew khat, thereby dramatically increasing the amount consumed domestically. As such, the rise of khat consumption was one of the distinctive socio-cultural transformations in Ethiopian society in the last half century.

The consequent rise in demand for the leaf was accompanied by dramatic increases in the land under khat cultivation and a gradual shift to cash cropping in the agrarian system. Official statistics of the total land under khat cultivation in Ethiopia are hardly accurate. The national figures tend to understate the land devoted to growing khat, apparently to avoid criticisms for an unrestrained growth of a crop many consider a deleterious drug. As conservative as the national figures are, the yearly rise shows the speed with which khat orchards have been growing in recent years. The official agricultural sample survey for the main harvest season of 2003 showed that khat was planted on 111,578 hectares of land (CSA, 2004).[2] In the subsequent years, the survey results showed 120,304 hectares (CSA, 2005) and 136,189 hectares under khat cultivation (CSA, 2006), nearly 8 percent and 13 percent increases over the previous year respectively. In 2006, it jumped to 147,805 hectares and peaked at 163, 227 hectares during the 2007-08 cycle, increasing by over 10 percent (CSA, 2006; CSA, 2008).

This has caused considerable concern in official circles and among development experts (Adinew, 2005; Tefera, Kirsten & Perret, 2003; Gebresellassie, 2006). Critics maintain that the high proportion of prime agricultural land devoted to the cultivation of khat, the amount of productive agricultural labor wasted on chew sessions, and the huge household expenditure on the leaves have retarded the country's economic development. They argue that khat obstructs the officially promulgated policy of greater national self-sufficiency in food production (Woldemichael, 2004). These and various other assumptions about the adverse effects of increased khat production require greater scrutiny than they have been given in the uninformed exchanges that have characterized the debate on khat in Ethiopia. Our knowledge – of the structure of khat production, the effects on agriculture of expanding cultivation, and the impact of the influx of high revenues from its sales on household income and the regional and national economies – is too inadequate to inform the formulation of sound agricultural policy (Stage & Rekve, 1998, p. 189). This chapter outlines the evolution of the salient features of the local khat economy and describes the economic effects of khat on agriculture, household income, and the local economy. It shows that khat has helped to promote and sustain agriculture and finance government operations. While the issue of khat is evidently a national concern, this chapter focuses on the Harerge highlands, the historic area where a shrub once produced for household consumption was transformed into a commodity traded all over the Horn of Africa.

THE REGIONAL ECONOMIC CONTEXT

Until the late 1980s, the khat economy was essentially an economic enclave whose impact was confined to the eastern part of the Harerge highlands. It has now become a mainstay of the national economy competing for land and labor everywhere in the country. Aside from strong external demand, there are internal reasons for the growing commoditization of khat and the concurrent expansion of its cultivation to nearly all regions of Ethiopia. One factor relates to a pattern of agricultural development that has yielded very low levels of income in the Harerge region, where the majority of the farms gradually fell in size from an average household holding of 2 hectares around the 1940s, to 1.5 hectares in the 1960s, to 1 hectare in the 1980s, and to 0.5 hectares in the 1990s (Wibaux, 1986; Adinew, 1991). Policy changes also played a major role in accelerating the pace at which farmers were forced to change their cropping patterns (Kassa et al., 2002). To a considerable extent, the physical characteristics and proximity to markets also conditioned the evolution of agriculture in the Harerge highlands, particularly in the eastern section of it (Wibaux, 1986; Gebissa, 1994; Gashaw, 1996).

The Harerge highlands may be divided into three major geographic regions. At the western end, the highlands are characterized by a complex surface configuration consisting of higher plateaux of undulating profile, featuring a few perennial streams flowing north through steep, narrow courses into the plains below. At the eastern end, the Harer or eastern highlands are characterized by long and terraced valleys of gentle slopes, which alternate between flat-topped mountains and elongated spurs. Along the southern slope, the Harer highland descends gradually into the Shebbelle Plains, its drainage carried through the four valleys of the Water-Gobellie, the Erer, the Dakata, and the Fafan rivers. Between the Chercher and Harer highlands lies a long spur that is projected south for a distance of about forty miles. This land mass forms the Gara Mula'ata massif, which is separated from the highlands by two deep valleys formed by two perennial rivers, the Watar in the east and the Ramis in the west. Because of its geographical location and altitude, the Harerge highlands normally receive relatively high rainfall, an estimated annual average of 785 mm at lower elevations to 1,200 mm for higher elevations (AUA, 1985; Wibaux, 1986; Mulatu, Ibrahim & Bekele, 2005).

The diverse agro-climatic conditions determine the composition of the rural economy. The main crops cultivated in Harerge highlands are cereals (sorghum, maize, barely, wheat, *teff*), cash crops (coffee, khat, Irish potato, sweet potato, onions) and pulses. Other corps, such as peanuts are grown in some lower elevations. Increased frequency of droughts and the variability and erratic distribution of rainfall even in relatively normal years have historically hampered the capacity of farmers to maintain a diversity of crops (Brooke, 1958). Coupled with an increasingly acute shortage of arable land, the margin of error in farmers' decisions that take account of agro-climatic conditions has progressively narrowed (Klingele, 1998). As a consequence, the agricultural system of the Harerge highlands has gradually shifted over the last half century, from a cereal-coffee-livestock integration, to a cereal-khat-vegetable complex, to an increasingly khat mono-crop regime (Mulatu, Ibrahim & Bekele, 2005, pp. 81-82; Kassa et al., 2002, pp. 101-103; Klingele 1998, pp. 17-18).

Another factor determining the nature of agriculture in the Harerge highlands is the region's strategic position between the Ethiopian interior and the coast of the Gulf of Aden. Historically, major trade routes connected the main commercial centers of the region with the rest of the world. The markets at Harer and Dire Dawa have been historically important as entrepôts for forwarding local produce, mainly cash-crops from the hinterland, to the coast and abroad as well as disseminating foreign commodities imported at the Gulf of Aden ports to consumers in the eastern highlands and further in the interior. Around the turn of the century, the region became the first to be connected by rail to the Djibouti port, substantially reducing the time it took to deliver goods to the ports (Gebissa, 2004, pp. 48-51). It also allowed for the export of bulk goods. This has endowed the region with lively markets and a vigorous system of networks that make international trade relatively efficient.

The historical importance of trade has allowed the Harerge highlands to have a relatively diversified economy, which combined subsistence and commercial agriculture, small-scale domestic livestock production, small trading businesses and transportation services. At different times, the relatively developed infrastructure of commerce has allowed the local economy to remain connected with the world economy. The enterprising traders of the region operated in the legal sector whenever the policy environment permitted and opted for the contraband sector when government controls became burdensome, thus creating a burgeoning commerce that thrived based on informal networks of credit and exchange. This situation has for more than a century allowed for greatly increased financial returns on agricultural commodities and more opportunities for employment locally, especially in the agricultural, commercial and transportation sectors that grew around the khat business (Gebissa 2004, pp. 181-182).

KHAT AS A CROP

There is no doubt that Harerge's commercial importance has played a critical role in integrating the region to external markets, but the prime movers of change in the agrarian sector are the region's population density and agricultural productivity. Because of high growth rate, population density in Harerge highlands has been very high: between 300 and 500 persons/sq. km (CSA 1999; CSA, 2005; Mulatu, Ibrahim & Bekele, 2005, p. 85; Brooke, 1959, p. 63). The size of the population relative to the available cultivable land has brought great pressure on land and made holdings relatively small, fragmented, and overused. In 1998, according to projections based on the 1994 census, a plurality of households — over 44 percent — owned small plots, ranging between 0.25 hectares and 0.75 hectares in area. About 15 percent of households had holdings between 0.75 hectares and 1.5 hectares land. Only about 3 percent of households had land between 1 and 3 hectares of land. Even fewer numbers of households possessed holdings more than 3 hectares, which was considered very large by local standards (CSA 1999). Today, the average household plot size has fallen below the commonly cited 0.5 hectares figures as the average holding size. For instance, in the Qarsa district of eastern Harerge, 43.1 percent of households own an average plot of less than 0.25 hectares whereas 7.5 percent of the households own between 0.25 and 0.5 hectares (CSA, 2008; Mulatu, Ibrahim & Bekele, 2005, pp. 87-88). At a growth rate of around 3 percent, researchers making projections in the late 1990s expected population to double in 20 years, further increasing pressure on land and other resources (Klingele 1998, p. 5). In fact, the scarcity of cultivable land began to push many farmers into the rank of the landless much sooner and caused a significant proportion of smallholders to devote their land to the cultivation of high value cash-crops or migrate to other regions (Piguet, 2002; Ahrens, 1998).

The reduction in size of household plots in fact started to be felt as early as the 1920s and farmers began to change their farming strategies in the 1940s. However, experts were praising the potential of Harerge's agriculture rather effusively even as late as the 1950s.

> The flourishing agriculture of the Central [Harerge] Highlands obscures the fact that the peaceful farm communities in the region represent a great transformation in the economic life of the people. Until 150 to 200 years ago nearly all of the Eastern [Oromo] were pastoral. ... An interesting aspect of agriculture in the Central Highlands is its relatively advanced techniques. The use of the plow, production and application of compost, knowledge of crop rotation, terracing, and irrigation are features which are not

widely found among cultivating peoples in Africa south of the Sahara (Brooke, 1958, p. 203).

At the time of this observation and in the subsequent three decades, farmers in the region continued to adjust their farm management practices and cropping patterns to cope with problems associated with shrinking land resources. The diminution of household plots and the resulting precariousness of Harerge's agriculture reached a critical mass in the 1980s as farmers became unable to sustain their families with income from an average of 0.5 hectares of family land. As more and more farmers shifted to high return cash crops, Harerge's agriculture became more commercialized, dominated by the production of khat. As a consequence, the Harerge highlands resorted to relying for much of its grain and other food requirements on the adjacent provinces or on food aid from NGOs (Piguet, 2003). This transformation occurred in four stages: 1) planting khat on marginal lands; 2) diversification through intercropping; 3) dramatic shift toward cash cropping; and 4) predominantly growing for market and importation of food crops.[3]

At every stage, we witness great reluctance on the part of farmers to abandon the principle of self-sufficiency in food production even as they steadily increase the production of cash crops in the face of dwindling farm resources. In the decades after the Second World War, the agrarian system in Harerge commenced a general trend toward a gradual expansion of the land committed to cash crop cultivation, even though the market for khat was small and episodic. The marketing system was entirely controlled by Yemeni Arabs who had the advantage of familiarity with the retail market in Aden and Djibouti. They kept prices low to increase their own margins. As Clarke Brooke (1958) observed in the late 1950s, "prices paid the farmer for his produce too often [were] unrealistically low in terms of a fair return to the cultivator and tend[ed] to inhibit his desire to produce beyond the subsistence level" (p. 204). This was a major factor in the low level of khat production and relatively small proportion of the land given

Crop and Commodity

over to its cultivation by the early 1960s (Getahun & Krikorian 1973, p. 367). Sorghum dominated the fields in Harerge, not only occupying "greater acreage than other cultivated plants in the region, but, without a doubt, its yield [exceeding] the combined total of all other grains" (Brooke, 1958, p. 193).

While self-sufficiency in food was a production imperative that guided farmers' cropping decisions, each farmer had an obvious need for cash that had to be met. Brooke (1958) noted: "Every farmer is obliged to sell some of his produce in order to pay his taxes and to buy necessary staples such as salt, cooking oil, and cotton cloth" (p. 204). In most places in Harerge, farmers supplemented their cash needs by growing cash crops, mainly coffee. Khat brought slightly better return per unit while requiring minimum input. Fortunately, that benefit came without threatening the farming principle of production for food self-sufficiency because, as Brooke (1960) noted,

> "khat is usually terraced, a technique which is not infrequently used on some of the steepest hillsides on the Harar plateau. ... Near the end of the rainy season-usually in September-shoots are cut from the root stock of mature khat and planted in the holes. ... When three years old the plant is about two and one-half feet high and gives a small commercial yield. Until the fifth or sixth year, maize, peppers, sweet potatoes, tobacco, onions, or sometimes even barley and *teff*, may be interplanted without undue shading by the young trees. Thereafter, other crops are rarely found in the khat area. ... Where there is sufficient flow, khat is irrigated every three to five weeks from November until the rains commence in March or April (p. 53).

At this stage, thus, farmers planted khat because it suited well with the famers' strategy of growing cash crops to meet their cash needs while remaining food self-sufficient. While experts at the time maintained an optimistic view that the commitment to food self-sufficiency will continue indefinitely, there was a palpable trend toward commercialization already at work. Clarke Brooke's (1958) assessment was almost prophetic.

> New or radical changes in the [rural] economy are not to be expected. The durra complex will continue to be dominant. But a greater measure of commercial agricultural production is sure to follow the gradual expansion of transportation facilities within the [Harerge] Highlands, which probably in time will bring the region into closer association with domestic and foreign markets (p. 204).

Planting khat as cash crop that Harerge's farmers began to undertake in the 1940s (Risoud, 1987, p. 38) continued in the 1950s at a gradual but measurable pace. Between 1954 and 1961, the size of arable land devoted to khat production in Harerge more than doubled, from 2,996 to 6,997 hectares (Table 5.1); and these figures presumably excluded those farms where khat was intercropped with other staples. Even as the size of the land under khat increased, the khat industry was of much less importance to Harerge's economy. The bulk of the khat that was produced then was for local consumption and was probably grown by farmers with larger holdings and occurred in areas with access to roads and urban centers where market opportunities existed. Contemporary photographs suggest that the entire Dhengego mountainside, about 20 kilometers from Dire Dawa, was bench-terraced and planted in khat. In the Haramaya area, khat was planted on sloping household plots to prevent erosion and raise extra income in the dry season. It seems that farmers used *marginal lands* for growing khat for market without having to jeopardize their ability to produce enough food for their families.

Only a small proportion of the total khat production was exported — mainly to Djibouti, the Somaliland Protectorate, and Aden (Brooke, 1960; Gebissa, 1994; Gashaw 1996). In fact, the volume of export dropped in 1964, because a serious border skirmish between Ethiopia and Somalia disrupted the trade and Aden decided to discontinue importation. In that year, the price of khat fell by up to 45 percent in the main wholesale market in Haramaya town (Getahun & Krikorian, 1973, p. 368). It is possible that this slowed the pace of plating new khat. By the late 1960s, however, the khat trade had recovered sufficiently that farmers resumed planting more khat. Getahun and Krikorian (1973) observed:

> There are many [khat] fields around the city of Harar, and there are extensive fields of [khat] in the [Kombolcha] and [Haramaya] areas as well. The entire upper part of Dengego Mountain is devoted to growing [khat]. One may occasionally see a field of sorghum, but it is actually a permanent [khat] field, as the practice is to interplant an annual crop such as sorghum. ....The entire area of [Haramaya], [Kombolcha], Harer and the eastern extremity of Chercher Highlands is hilly, and erosion is sever, so [khat] is grown on the hillsides, which are usually terraced, while lower (fertile) lands are devoted to sorghum, corn, vegetables and grazing (Getahun & Krikorian, 1973, p. 358).

It is essential to note that, despite the relative profitability of growing khat, farmers had not abandoned their commitment to food crop production by the early 1970s. The descriptions offered by Clarke Brooke in 1960 and by Getahun and Krikorian in the early 1970s of the Haramaya-Kombolcha region were strikingly similar. Khat was being planted on hillsides which, when bench terraces were built, opened up more land for interplanting with food crops (Brooke, 1960, p. 53). Khat planting produced a net gain for the farmer since it maximized income by using land that would otherwise be left to erosion. This evolution of the agrarian system was not quite evident in the Chercher highlands, which lagged behind the Harer highlands in terms of the pace of shifting to khat (Strock et al., 1991).

FIGURE 5.1: TERRACED MOUNTAINSIDE KHAT PLANTATIONS, UPPER DHENGEGO, c. 1970.

From Getahun & Krikorian (1973).

By the early 1970s, khat had become an important part of the cropping system, competing for land with coffee. Getahun and Krikorian (1973) observed: "[Khat] culture seems to be gaining momentum in its spread, and in many areas where coffee was once grown, [khat] is now preferred since the net return per acre is greater than that from coffee" (p. 357). The agricultural sample survey conducted in 1975 showed that khat covered 26,800 hectares or 6.6 percent of the cultivated land, increasing by 283 percent over 1961. Khat was displacing coffee because of its relative profitability (CSA, 1975). The income from khat per unit was ten times larger than the income from coffee (Getahun & Krikorian, 1973, p. 367). Even so, the logic of farming for the majority of Harerge farmers reflected a strategy of *diversification through intercropping* rather than an unswerving commitment to cash crop expansion at the expense of food crops.

Events in the mid-1970s forced farmers to reconsider their strategy of maintaining food self-sufficiency through diversifica-

FIGURE 5.2: TERRACED MOUNTAINSIDE KHAT PLANTATIONS, LOWER DHENGEGO, c. 1970.

From Getahun & Krikorian (1973).

tion. In 1974, Ethiopia experienced a revolutionary upheaval that ended imperial rule. The Derg or the military junta that came to power implemented new agricultural policies that played a major role in transforming the agrarian economy. Driven by a determination to distribute land to smallholder farmers and to end the problems of evictions and landlessness, the new government initially favored 'smallholder production' by farmers with free tenure and government provision of essential services. The initial commitment to freehold smallholder production gradually gave

FIGURE 5.3: KHAT CULTIVATION AREAS AND REGIONAL STATES, 1990s

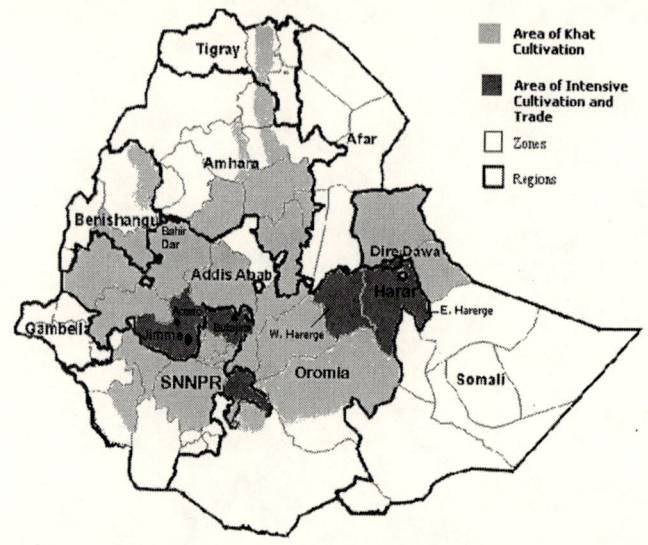

way to a state-controlled, centrally-guided socialist modernization of the country's agriculture. The elements of the new policy included the introduction of price controls, development of state-controlled mechanized farms, and expansion of the role of the state in agricultural marketing, specifically coffee. The coercive approach with which the policies were implemented convinced smallholders that government-sponsored avenues of agricultural development were not in their best interest. They came to view the policy as an ideological drive to control market structures and to privilege state-controlled producers' cooperatives. In response, farmers increasingly migrated to the production of a relatively high value cash crop that was less regulated, khat. Many farmers uprooted their coffee trees to make room for khat (Gudeta & Kahssay 1998, pp. 5-6; Anderson et al., 2007, p. 26).

The result was dramatic. In 1975, khat was planted on 6.6 percent of the total land under cultivation in Harerge. In less than a decade, this proportion had doubled and, by 1983, khat

TABLE 5.1 PRODUCTION: LAND USE TREND IN HARERGE, 1954-2000

	Land under khat cultivation (hectares)	% Change	% of total cultivated land	% of land under coffee
1954	2,996[a]			
1961	6,997[a]	133		
1975	26,800[b]	283	6.6	6.6
1983	96,445[c]	259	13	4.4
2000	112,206[d]	16	14	6.1

Source: a. Hagos, 1963. b. CSA 1975; c. EEPO, 1988; d. EFSC, 2001[4]

accounted for 13 percent or 96,445 hectares of the cultivated land. Khat's main rival cash crop, coffee, which occupied 6.6 percent of the area under cultivation in 1975, dropped to 32,643 hectares or 4.4 percent of the total cultivated land. Khat, a crop planted on marginal lands only two decades before to supplement farmers' cash income, surpassed coffee as a cash crop of choice and was planted on land previously used for growing coffee (Gebissa, 2004). Khat consumption and cultivation increased in a dialectical relationship throughout the 1980s and the shift to khat became more conspicuous on the land. Concerning the transformation of the agrarian system, Risoud (1987), a researcher at the Haramaya University Farming System Unit in the mid-1980s, stated: "One of the main transformations of the agrarian landscape was, from the end of the 1940's to the beginning of the 80's, the overall augmentation of the area cultivated with [khat]" (p. 38).

Even though the political environment under the Derg was unfavorable to growing khat because of the government's preference for coffee, and the increasingly uncertain rain patterns in the 1980s, planting khat was employed by more and more farmers as a strategy to maximize income from a rapidly shrinking household plot. An important difference from the imperial period was that farmers held land in usufruct and were to some extent able to make cropping decisions. In areas where soil conditions allowed and irrigation was possible, such as the Haramaya and Kombolcha areas of the Harer highlands, the cultivation of khat increased dramatically. The guiding principle in farmer decision making was food security through an efficient use of land characterized by *a dramatic shift to cash cropping*. By the end of the 1980s when the Derg was losing control of its power, farmers' traditional reluctance to abandon food crops was apparently being replaced by a commitment to finding the right balance between high-income crops and growing food crops to ensure household food security. Speaking of the situation in Harerge, Risoud (1987, p. 38) noted:

> ... the farmers simultaneously operated a transformation of the existing cropping patterns toward cash crops [khat mainly] and an intensification of the productive practices with the development of association of crops. From an analytical point of view, this corresponds to a trading strategy as well as food self sufficiency at the level of the family farm.

The expansion of khat cultivation was a sign of a more profound shift in farmers' strategy to cope with political and environmental challenges. In the 1990s, the pace of planting khat continued to accelerate. In 2000, according to one local report, citing the regional office of the Ministry of Agriculture, khat was planted on 14 percent of the total cultivated land or 98,837 hectares in the Eastern Oromia Zone; i.e., eastern and western Harerge. In 2001, the figure jumped to 106,722 hectares in the same Zone, increasing by 7,885 hectares or 7.9 percent over 2000. In the same year, the land committed to khat in Eastern Ethiopia was 112,256 hectares (Export Facilitation and Support Committee EFSC, 2001). This time, khat trees were standing on land previously used for food crop production. More significantly, because the khat plants were large trees, they required more manure and water, making intercropping more difficult.

TABLE 5.2 KHAT CULTIVATED LAND AND PRODUCTION LEVELS, EASTERN ETHIOPIA, 2000

Administrative Unit	Cultivated Land (hectares)	%	Production (kg/hectare)*	Production (tons)	%
Eastern Harerge Zone	77,750	69.2	1250	97,186	72.2
Western Harerge Zone	28,872	25.8	1100	31,759	23.6
Hareri National State	2,664	2.4	1250	3,330.	2.4
Somali National State	2550	2.3	800	2,040	1.5
Dire Dawa Council	370	0.3	800	296	0.2
Total	112,206	100		134,612	100

Source: EFSC, 2001; FAO, 1984. Estimate of 1000kg of green khat harvest per year. Wakjira (2004) established 1500 kg/hectare for the Haramaya (Eastern Harerge) area. Western Harerge yield per hectare/year averages between the two estimates.

TABLE 5.3 CROPPING PATTERNS IN HARAMAYA, 1991-2003

Crop	Pre-market liberalization (1991)		Post-market liberalization (2003)		Change	
	Area (hectares)	% of total	Area	% of total	Absolute	Percentage
Sorghum	62.4	65	20.00	20.83	-41.57	-66.62
Maize	18.24	19	13.00	13.54	-5.24	-28.73
Khat	7.68	8	52.01	54.18	144.33	577.21
Vegetable	6.72	7	9.79	10.30	3.07	45.69
Others	0.96	1	1.2	1.15	.24	25.00

Source: Adopted from Wakjira (2004).

Yield per hectare varied from place to place in the Harerge highlands. Some studies estimated average yield per hectare to be 1000 kg of fresh khat (FAO, 1984). Recent studies have shown that 1500 kg per hectare is possible under favorable soil and moisture conditions and appropriate management (Wakjira, 2004). According to Klingele (1998), the average yield per hectare is between 700 and 1000 kg. All the estimates for Harerge highlands are considerably less than the aggregate reports shown in national figures (CSA, 2004) or reports of between 3321 kg and 6235 kg per hectare in Yemen (Sherif et al., 2002). Given the range, it is not unreasonable to assume 1250 kg per hectare per year for the Harer highlands, 1100 kg per hectare for the Chercher highlands, and 800 kg for the rest of the khat producing districts of the Harerge highlands. Based on these figures, the annual volume of production of khat in the eastern highlands is computed to fall between a low total average of 89,764.8 tons and a high total average of 140,257.5 tons. Calculated based on estimated average yield per hectare for each region, the estimated total annual production of khat in eastern Ethiopia was 134,612 tons in 2000.

By the early years of the present decade, while food crop and coffee production still constituted a significant portion of the cropping regime in most parts of Western Harerge, the shift toward cash cropping seemed nearly complete in some districts in Eastern Harerge. The fall of the Derg in 1991 and the subsequent policy reform, particularly the abolition of state farms and peasant cooperatives as well as the lifting of price controls, allowed farmers to respond to market incentives and expand their cultivation. At the same time, khat consumption increased substantially and with it the price per unit for khat. For example, the price of exported khat, which was 11.02 birr per kg in 1980/81 rose to 20.21 birr in 1991/92, and reached 33.95 birr in 1992/93 (Gudeta & Kahssay, 1998). By 2003, price had risen to 81.54 birr and 96.98 birr the next year, owing to the beginning of legal export to Hargiessa (NBE, 2004). Farmers responded by increasing the amount of khat production. According to a survey of 120 households in Haramaya district, land under khat cultivation rose from 7.68 hectares in 1990/91 to 54.18 hectares in 2003/04, increasing by an astonishing 577.21 percent, while the acreage under staple crops, sorghum and maize, dropped by 66.62 percent and 28.73 percent respectively (Wakjira, 2004).

Table 5.3 shows that vegetable production has been increasing at a relatively moderate rate than khat. Other studies show that, in some areas of the Haramaya district, vegetable production has been rising slightly faster than khat (Gudeta & Kahssay, 1998). Either way, Haramaya has in recent years experienced an accelerated pace of agricultural transformation, with a visible shift to *predominantly producing for market*. It seems price exerts a strong influence in determining cropping patterns and farmers respond by adopting a strategy of a combination of crops that would give them maximum income. Even in the economic environment where cash crops fetch substantially higher income, Harerge's farmers never gave up producing food crops for household consumption. This may be the result of limitation of resources such as water for irrigation and fertilizer to wholly focus on cash crops or administrative action which generally tends to encourage food crop production (Stage & Revre, 1998, p. 191). At any rate, there is no denying that the logic of farming which has characterized Harerge's agriculture for a long time, the commitment to food self-sufficiency, is under severe strain. Agricultural economists estimate that it takes between 1 and 2 hectares of well-watered land to yield grain sufficient to support the average family in Harerge, which is 6.6 persons or one six-person family (Mulatu, Ibrahim & Bekele, 2005). The vast majority of families in Harerge do not have the potential to be self-supporting in grain production. Given that the community can never be smaller than it is today, it must be expected that Harerge will have to rely on the surrounding surplus producing areas such as the Arsi region for its grain requirements (Hailu, 2005).

Even though farmers in eastern Harerge desire to be food self-sufficient, the reality of the situation in their region does not allow them to maintain a commitment to crop diversification that some observers believe is still underway (Hailu, 2005; Stage & Revre, 1998; Tefera et al., 2004). The principle, if not the practice, of diversification in the form of cereal-livestock-cash crop integration might work in areas of Hararge where arable land shortage is relatively less acute. This cannot be said of eastern Hararge, particularly the Haramaya-Komolcha region. According to one study (Save the Children Fund (UK)–Ethiopia, 1996) which classifies Harerge into three economic systems, Haramaya is now a cash crop major system (distinct from the cereal major system and cereal and cash crop major system) in which khat dominates. Given that the shift toward khat begun in the Harer-Haramaya-Kombolcha axis, the trend line for the whole of Harerge highlands clearly points to graduated stages towards the cash crop major system and increasing reliance on food importation from food surplus areas. At the rate population is growing in this region and the irony of depending on the income generation potential of a shrub some consider a harmful drug, as Klingele (1998) observed some years ago, "the future of agriculture looks very bleak."

KHAT AS A COMMODITY

A key feature of khat trade is the leaf's perishability. Because cathinone is unstable, the commodity requires a highly efficient network to deliver it to consumers before it disintegrates to the less potent cathine. A large market network makes possible the delivery of khat from the farm gate to the consumer's cheek (Carrier, 2005). In general arriving at the market early is likely to ensure higher price and a handsome profit. However, traders cannot expect to reap high turnover by increasing volume on the assumption that yesterday's price would hold. Doing so might flood the market with excess supply that could further drive prices down or even increase the risk of a huge loss because wilted khat has little or no demand at all. Market uncertainty could be reduced by the growing speed of information relay, courtesy of mobile phones, but there is no evidence that the latest technologies have made the khat market more efficient thus far. There are major disincentives facing any khat trader who might have the capital and inclination to expand the size and scale of his enterprise. The very nature of the trade militates against speculation, market manipulation, and monopolization by big merchants.

A related characteristic of the khat trade is the absence of market sophistication that would otherwise keep low skilled individuals at bay. Khat marketing requires no special skills, large start-up capital, or custom-designed vending space. The market is largely unregulated and transactions often take place in open markets. The retailers, the majority of whom are unlicensed women vendors who usually operate without income tax obligations, buy their khat directly from farmers and supply poor quality khat to urban chewers in small bundles called *aqara* weighing approximately 1.50kg. The wholesale buyers, usually agents of a licensed exporter, are apprentices who had made their way up the career ladder. To make sure the leaves remain succulent, they buy whole crops on the tree and hire their own workers when the leaves and twigs are ready for harvest (Wakjira, 2004; Gebissa, 2004; Gashaw, 1996). At other times, they subcontract the buying to agents who have special relationships with khat farmers or agents of farmers from Badessa, Qunni, Habro, and Chiro in Western Harerge. The khat is trimmed, sorted and wrapped in smaller bunches of different varieties (*uratta*, *qarti*, and *hamerkot* etc) which are then bundled together in larger bunches of approximately 2.20 kg called *ferdo*. These activities require no prior skill to enter the industry. With increases in production and consumption, the number of people working in the khat industry has increased considerably. Even though four export companies were the principal exporters of khat in the early 2000s, they have little sway over the domestic market, which is the province of numerous individuals engaged in a variety of khat marketing activities (Anderson et al., 2007, pp. 45-47; Wakjira, 2004).

As a an agricultural commodity, the value of khat is determined by such factors as agro-ecological situations, farming practices, seasonality, supply and demand, risks, and skills at price negotiations. Because khat is a luxury commodity, the material qualities of the leaf may also influence consumer preference and hence price. Khat produced in Haramaya and Kombolcha has maintained price advantage for decades, despite the proximity of the khat fields to Harerge's main wholesale market at Aweday and the main retail markets in Harer and Dire Dawa. This may have to do with the reputation of the leaves as having a pleasant taste and potency. The variety from Badessa district in Western Harerge, though preferred in Djibouti, has not caught up in terms of popularity with the Haramaya one locally and in Somalia. The bulk of exported khat comes from Badessa while the Haramaya khat is used for blending and improving flavor.

The informal nature of the market does not generate data on the price situation of the domestic market that allows for mapping the price trajectory over a long period. Calculation of khat prices based on export figures shows a steady increase in khat prices over the last four decades. During the five years before the Ethiopian revolution of 1974, exported khat prices averaged 2.61 birr per kg. After the revolution, price rose to 2.86 birr per kg and then shot up to 8.28 birr per kg. In the early 1980s, when the Derg regime at once embarked on eradicating khat and monopolizing the export market by establishing a single, government-affiliated export company, price increased considerably, reaching 14.12 birr per kg in 1989. The lifting of price controls after the fall of the Derg occasioned further price rise, to 20.21 birr per kg in 1992, and steady increases thereafter (NBE, 2004).

TABLE 5.4 CHANGES IN THE RETAIL PRICE OF KHAT PER KG, 1970s – 2004

Variety	Surplus			Lean		
	Pre-1974	*1974-91*	*2003/04*	*Pre-1974*	*1974-91*	*2003/04*
Badessa (*Uratta diimaa*)	0.25	1.50	12.00	0.35	3.00	35.00
Haramaya (*uretta daalota*)	0.50	10.00	60.00	1.50	45.00	200.00
Other	0.20	1.15	15.00	1.70	15.00	40.00
Total	0.95	12.65	87.00	3.35	63.00	275.00
Average	0.30	4.22	29.00	1.18	27.00	91.66

Source: Adopted from Wakjira (2004)

The export figures reflect average prices at the Dire Dawa airport. At the marketplace, prices vary based on quality and the area where the khat was harvested. Much of the khat exported by companies comes from Badessa, which usually fetches considerably lower price than the khat from Haramaya or Komobolcha areas. Milkessa Wakjira's (2004) calculation based on internal revenue records and surveys of current prices shows the retail prices of khat increasing steadily since the early 1970s. Between the months of April and October, when the rains are relatively high, good quality khat (*uratta*) from Badessa sold on average for 0.25 birr per kg before 1974 and maintained an average price of 1.50 birr during the Derg period. The price for Haramaya khat was double the Badessa variety before the fall of the imperial regime, nearly seven times higher during the Derg years, and fivefold higher in 2004 over the Derg period. During the rest of the year, all varieties of khat were sold for slightly higher price both before and after 1974. Haramaya's khat, owing to high demand during the lean period, sold for about fifteen times more than the khat from Badessa. In 2003/04, the average khat price in Harerge averaged 29.00 birr per kg during the surplus season and 91.66 birr per kg during the lean season. Prices were much higher for Haramaya and Kombolcha khat, mainly exported by private individuals numbering about 100, averaging about 70 birr per kg and rising to between 300 and 500 birr in lean seasons (Wakjira, 2004).

A large sum of money, generated by the khat trade, circulates in the region. In 1994, for instance, an estimated 225 buyers came to the evening market and 350 persons to the early morning market at Aweday, Harerge's largest market center. Each buyer collected various quantities of khat, ranging between 60 and 700 kg. Based on these figures, an average of 218,500 birr ($43,720 in 1994) worth of khat changed hands at the marketplace every day. This figure translates to 79.8 million birr or $16 million spent annually on purchasing khat only in Aweday (Gebissa, 2004). In 2002, an estimated 25,000 kilos of khat were sold at the Aweday market every day for approximately 375,000 birr (Bhalla, 2002). One recent study (Anderson et al., 2007) reported that, in 2004, no fewer than 5000 persons visited the Aweday market "doing business in khat valued at 10 million birr every day" (pp. 44-45). Indeed a lot of money circulates in Aweday, but an estimate of 3.65 billion birr annually is a bit exaggerated even for a market described as the second largest marketplace in Ethiopia, after the *marcato* in Addis Ababa. In any case, khat-generated money is large and bound to have some effect on the household incomes of those who are engaged in khat production.

ECONOMIC EFFECTS OF KHAT

HOUSEHOLD INCOME AND FOOD SECURITY

Growing khat not only generates substantial revenues for the country and considerable wealth for everyone associated with the industry. If as some studies suggest 3,000 khat trees are planted to the hectare, the yield would be between 1500 and 2,000 *aqara* of khat a year (EFSC, 2001; Gashaw, 1997, p. 60). In a year when the average price of an *aqara* of khat in the market is high, the market value of the khat from 1 hectare of land falls between 142,500 birr (approx. $18,000) and 190,000 birr (approx. $24,000). If we assume the average farmer planted khat on household average land size of 0.5 hectares, the market value of the khat on that land in 2003 would have been about 20,000 birr (approx. $2,500). These figures indicate the total cash revenue generated from the average plot and do not represent the farmer's gross income. When the cost of production is accounted for and benefits are allocated to all the parties involved, the farmer is left to a fraction of the total cash value. If we assumed the farmer's gross income to be about 25 percent of the total value, that would mean a gross margin of 5000 birr ($625) in 2003. That is well above the per capita income of $110 for the country adjusted for inflation and purchasing power for the same year (World Bank, 2004).

Income data in a subsistence farming setting are hard to come by and whatever that exists is not reliable. As a result, many of the studies on the impact of cash cropping tend to focus on smaller units such as a district or a village where the researcher can generate data. A few studies that have looked at the entire Harerge highlands region have consistently shown that the cash income of khat farmers is at least three times that of cereal growers (Getahun & Krikorian 1973; Wibaux, 1986; Klingele, 1998; Mulatu & Kassa, 2001; Gebissa, 2004). There is now consensus among scholars that khat, as a cash crop, has had a positive impact on the quality of life of producers, traders, and others who earn their income from activities connected to khat (Seyoum et al., n.d.; Tefera, Kirsten & Perret, 2003).

Describing the improved quality of life khat has afforded ordinary farmers, Ahmedin Muktar, a khat trader from Harerge, commented: "If you compare khat farmers with other farmers, you will see that their standard of living is so much better. They have good houses and some even have cars. How many farmers do you know who have cars?" (Bhalla, 2002, para. 9). The benefits of khat are not limited to the male farmer. Alfiya Ali, a 36 years old housewife from the Wolembu village in Haramaya, describes with a sense of contentment the improved life khat has afforded her as follows.

> [We] supplement the family food requirement with some rice, spaghetti, macaroni, wheat flour and sweet potatoes bought from the market. The expense for these items too was mostly born [sic] by [khat] sale. ... The daily cash demand of the household for other expenses (cooking oil, gasoline, salt, sugar, fenugreek, and potatoes), which amounted to 2.5 Birr a day was secured from [khat] sale by and large (Quoted in Gashaw, 1997).

In addition, families reabsorbed people who could not succeed in cities, hiring them in such jobs as drivers for their commercial vehicles. Farmers generally radiated self-confidence, although some were unsure of the sustainability of the market for the leaf (Gebissa, 2004). The majority of farmers benefited from the high price of their khat, the only commodity whose price kept ahead of inflation and that, during the command economy years of the late 1970s and 1980s, evaded government price controls (Table 5.4).

Empirical studies confirm the testimonies of farmers and the observation of scholars. One study, focusing on the socio-economic impact of export crops from eastern Ethiopia on the livelihood of the farmer, reveals that producers of agricultural exports, particularly khat producers, are better off than non-producers in terms of education, food security, and housing conditions (Kedir, 2005). Using mean test and regression analysis, the study shows 66.5 percent of producers as compared 31.2 percent to nonproducers of export products sent their children to primary school, indicating sufficient cash to cover school expenses. Moreover, 70.2 percent of producers as opposed to 29.8 percent of nonproducers are shown to have the ability to feed their children more than once a day. Similar percentages held true for the frequency of adults being fed. The study also showed that producers were better positioned to withstand shocks in times of food shortage, crop failure and/or other difficulties as evidenced by the high proportion (45.2 percent) of nonproducers relying on food aid to supplement food production compared to producers (23.1 percent). Because khat producers are more able to withstand shocks, it is possible to state that increased cash cropping has made them more food-secure than the non-producers (Stage & Revre, 1998).

In addition, increased cash cropping has improved the living conditions of farmers. A higher percentage of producers own houses roofed with corrugated iron sheeting and have separate kitchens for cooking and separate structures for animals. The distribution of farmers having corrugated iron roofed houses is about 49 percent for producers to 26 percent for non-producers. The percentage of farmers using separate kitchens for cooking other than their living rooms is also significantly higher, about 41 percent for producers and 22 percent nonproducers. Those who maintain a separate structure for animals is slightly lower for producers (37 percent) than nonproducers (39.5 percent). Overall, the underlying fact is that families that produce export agricultural commodities are more likely to send their children to school, be more food secure, and have improved dwelling conditions than the non-producers (Kedir, 2005).

Another study, specifically focusing on khat farmers, shows similar patterns. Using proxy indicators such as livestock ownership, value of farm implements, expenditure and ownership of houses with corrugated iron sheet roofs, or ownership of valuables, it shows that khat growers are significantly better off than the nongrowers in terms of their household income (Tefera, Kirsten & Perret, 2003). With regard to the issue of whether increased income from khat production is translated into improved food security and nutritional status, it is important to state that food security has several interrelated components, including expenditure behavior, gender relations, availability, and price of food. The available data are not adequate to account for all these factors, but all indications are that khat growers are not just more food secure but also more food self-sufficient than nongrowers (Table 5.6).

TABLE 5.5 COMPARISON OF INCOME FOR KHAT GROWERS AND NON-KHAT GROWERS IN BIRR

Proxy indicators	Khat growers	Non-khat growers	%
Cash crop income	2499.95	444.84	18
Total expenditure	2506.95	1226.57	49
Value of farm implements	914.62	223.51	24
Livestock owned in typical livestock unit	2.72	1.70	62
Percent with iron-sheet covered house	58.7	40.6	

Note. From Market Incentives, Farmer's Response and a Policy Dilemma: A Case Study of Chat Production in the Eastern Ethiopian Highlands, by T.Tefera, L. Kirsten & S. Perret, *Agrekon* 42: 3 (2003).

According to the data in this study, increased khat production has little impact on the nutritional status of farming households, but the frequency of feeding or the number of times children are fed each day has been shown to be higher with khat producers than nonproducers. Interestingly, the frequency increased commensurate with increases in the land devoted to khat production and decreased as land allotted to cereal cultivation increased. While several other factors may be involved, it is certain that growing a cash crop does not necessarily mean food insecurity or deficiency. As Mulatu and Kassa (2001) rightly observed, "the role of khat in improving household food security of the rural poor in the Harer Highlands cannot be over emphasized" (p. 107).

TABLE 5.6 FOOD SECURITY INDICATORS FOR KHAT AND NON-KHAT GROWERS

Indicator category	Khat growers (%)	Non-khat growers (%)
Production surplus	15.2	11.4
Self-sufficient	25.6	11.4
Self-insufficient	59.2	77.1
Never faced food shortage	44.8	35.7
One year shortage	17.6	17.1
Two years food shortage	19.2	18.6
Shortage of three to five years	18.4	28.6
Don't sell grain	56	44.3
Net grain per adult available for consumption (kg)	255.2	214.2

Note. Adapted from Market Incentives, Farmer's Response and a Policy Dilemma: A Case Study of Chat Production in the Eastern Ethiopian Highlands, by T. Tefera, L. Kirsten & S. Perret, *Agrekon* 42: 3 (2003).

In addition to the improved cash income and its effects on food security, khat has positive agronomic consequences. The growing integration of khat into the farming system has perhaps benefited farmers directly by providing a temporary solution for the apparently intractable problem facing Harerge's agriculture—shortage of arable land. In Harerge, particularly in the eastern and central districts, as much as 90 percent of the land is cultivated and the average household land size had dwindled to 0.84 hectares in 1990 compared to 1.1 hectares in 1980 and 1.5 hectares in 1965 (Bishaw, 1993). Recently, studies have shown that the average household size had further declined to less than 0.5 hectares (Mulatu, Ibrahim & Bekele, 2005, p. 87). In the 1980s, the average yield per hectare for sorghum was 19 quintals. Since it is impossible to survive on this quantity of sorghum, farming households must maximize the productivity of the available land. No other crop that the farmers adopted kept the price differential and profit margins of khat.

By the early 1990s khat had become part of the diversified cropping system that fit well into both the environment and the agricultural cycle of small-farm households without being too demanding of labor (Mulatu & Kassa, 2001, p. 99). Harvested at least three times a year, it provided a regular source of income to meet household expenditure. Khat also consistently maintained price advantages over other crops. In the 1980s, it yielded three times the returns on sorghum, 10 times the returns on coffee, and nearly twice the returns on low input management potato. By the mid-1990s, the margin from khat was 20 times that from sorghum or maize. This means, the khat farmer could purchase twenty times more food crops by growing khat. In recent years, this price differential had widened further, with per hectare profit from khat exceeding all other crops,[5] making government-sponsored substitution programs unattractive and unsuccessful.

Government officials and NGOs still express concern that khat growing is taking over coffee plantations and food cropland in some parts of the Harerge highlands and the other regions of Ethiopia. Despite the abolition of the draconian means of controlling the coffee market during the Derg period, the value of Ethiopian coffee has continued to decline because of competition from new coffee producing countries. Since the profits from coffee for the farmer have remained a fraction of those from khat, coffee cannot continue to be cultivated entirely for financial reasons. However, there are still important incentives for cultivating coffee, although they are probably more social than economic. Coffee remains an important crop in the Western Harerge highlands. The farmers have always had a diverse economy; most of them derive their incomes from cultivating a range of crops and engaging in more than one occupation. Farmers are also aware that the demand for khat and high prices are linked to sustained demand from a consumer market. Only recently has Harerge's khat found new markets in Europe and North America, but even that is only a niche market among recent immigrants. Many farmers are dubious that new and reliable markets could be found. It is therefore likely that they are conscious of the prudence of agricultural and economic diversification. Keeping some coffee trees is not just a commitment to diversification, but also an insurance against a catastrophic event that would damage the khat market. This fact makes the farmers in Chercher reluctant to give up completely the coffee tree that takes six years to give its first harvest. Unlike the Harer highlands, where the pressure on land is more acute, the Chercher highlands remain a cereal-cash crop major system, with coffee as an important cash crop.

The concern that the trend towards cash cropping will inevitably result in decreased production of food crops, leading to malnutrition, disease, and food insecurity is not borne out by the realities of life at the farm level. Abrahim Hasan, a farmer from eastern Harerge, asserts that khat cultivation actually increases food production. He relates: "The life of the society and that of other crops depends on the life of [khat]. We would not have tended other crops except for the application of modern fertilizers which are only available through [khat]-made cash" (Quoted in Gashaw, 1996, p. 49). This is not simply a view of Harerge's farmers or a phenomenon unique to eastern Harerge. Examples of the synergy between cash crop and food crop production in other contexts abound. In Zimbabwe, for instance, "households engaging intensively in cotton production obtain higher grain yields than non-cotton and marginal cotton producers there" (Govereh & Jayne, 2003). In the Sudan, studies have likewise shown that revenues from cash cropping are generally used to finance food production. Traditional small farmers who produce more cash crops for the market are those who are more able to use purchased inputs to produce food grain (Elamin et al., 2003). In the French speaking parts of Western and Central Africa "the combination of cash crops for sale and food crops for consumption by the farmer and his family has produced very positive results in terms of increasing yields for all crop types" (Rafn, 2002). The fear that khat is contributing to reductions in food crop production in Harerge is, to use the words of Hamilton and Fisher (2003), a classic case of the disconnect between the mostly positive perceptions of farmers about their situation and the negative assessments of many who study non-traditional export agriculture.

The most enduring impact of khat is the fact that its production provided much-needed capital enabling individuals to leave their farms and look for nonagricultural opportunities elsewhere. The transport businesses, brick factories, filling stations, and small shops that sprouted all over the Harerge highlands were the result of khat-generated capital investment. These ventures have enabled farmers to extricate themselves from arable land that had become extremely scarce and had thus diminished the possibility of their self-sufficiency. In this regard, khat did what policy planners would have never imagined they could do for farmers: move them to nonagricultural occupations, thereby easing demographic pressure on land and agricultural resources. It is interesting to note that farmers pursued diversified occupational strategies to cope with the challenges of agriculture, whereas policymakers sought to fix the existing problems of agriculture and perpetuate smallholder subsistence farming.

REGIONAL AND NATIONAL IMPACT

During the imperial period, revenues from khat taxes financed the extension of formal administrative structures into the Ogaden region. During the Derg period, the effect of the social dislocation caused by the Ethio-Somali War of 1977 was eased by allowing

Harerge residents to engage freely in khat marketing and related business in contraband goods (Gebissa, 2004, p. 177). Spin-off benefits of the khat business extend to companies such as Ethiopian Airlines, which transports the leaf to international destinations, and local businesses in Harer and Dire Dawa that provide employment to thousands of the urban unemployed.

As I have shown elsewhere (Gebissa, 2004), the dramatic rise of demand for khat in Somalia, Djibouti, and Yemen after the Second World War led to a steady growth in Ethiopia's khat export. By the time of the fall of the Derg regime in 1991, khat has become a critical export item for Ethiopia (Ethiopian Customs and Excise Authority [ECEA], 1961–1991). Between September 1999 and August 2000, 15,684 tons of khat was exported and earned 619 million birr ($72 million), increasing by a staggering 84 percent in volume and 65 percent in value over the same period of the previous year. This figure amounts to about 14 percent of Ethiopia's export earnings and perhaps constitutes the second largest foreign exchange earner for the country (Moresh, 2000; Feyissa & Aune, 2003, p. 186).[7] In 2003/04, the country exported 17,825 tons of khat and earned 758. 9 million birr, ranking seventh among all commodity groups in volume but maintaining its second position in value (NBE, 2004). In the Ethiopian fiscal year that ended on 8 July 2008, Ethiopia exported 22,390 tons of khat to several countries, earning $ 108.3 million (approximately 965 million birr) in foreign exchange (Walta Information Center quoting Ethiopian Foreign Trade Promotion official, 2008).

In addition to the considerable export earnings, the domestic market generates substantial revenues for state and local governments. Under the imperial regime, taxes were collected at makeshift tax booth set up at major tax markets and inspection points along the main roads leading to border areas with the adjacent countries where khat was consumed. The Derg government introduced a system of collection at tollgates along main highways, located usually at the entrance of major khat consumption centers or market places like Dire Dawa, Harer, Aweday, and Jijiga towns.

Under this system, which the current government has retained, taxes were levied based on the estimated sales value regardless of whether khat was exported or sold in retail markets to consumers. The amount of taxes levied is assessed on each kg of khat that enters the market centers. Officials recognize the need for speed and do not require traders to make formal declarations. The individual assessor at the tollgate determines the amount of tax that is due. Since 1991 taxes have steadily risen due to the government's declared desire to finance economic development by increasing domestic revenue though efficient collection. In this context, it is difficult to overstate the importance of khat revenue and the government's determination to collect it more effectively. At the state level, tax revenue figures from 1995 to 2004 of the Oromia regional government, the largest state government and the main khat-producing region, shows an upward trend with revenues reaching 147 million birr in 2004. The share of tax revenue from khat as a proportion of the region's total income tax revenue increased from a little over 27 percent in 1995 to nearly 29 percent in 2004 (Arrafaine, 2004).

Projections based on export figures show that khat's contribution to the national revenue pool has grown steadily over the last four decades, increasing more than ten-fold between 1965 and 2004.[7] Calculated on the basis of the widely accepted view that the proportion of exported khat from eastern Ethiopia has never exceeded 10 percent of the total production, an average total of 15,350 tons of khat was marketed in the Harerge highlands during the last few years of the imperial government. The figure rose to 23,360 tons during the Derg period, to 31,590 tons in the early years of the current government, and 126,650 tons between 1999 and 2004 (Table 5.7). Actual export figures show a sharp rise in tonnage in 1999, reflecting the resumption of official export to Hargiessa in 1998 more than three decades after it was banned in 1964.

Analysts estimate that farmers market on average 60 percent of their total production and consume the rest. This means, khat generated an average of 1.4 million birr in revenue annually during the last decade of the imperial period and an average of

5.6 million during the Derg period (CSA, 2003b; CSA, 2004). The current government not only lifted the various restrictions that the Derg government had imposed on khat production and trade, it also raised the tax rate at the same time. The administrative restructuring that occurred following the establishment of a federal administrative system in 1995 led to the proliferation of tax collecting authorities. A khat trader described this to Ralph Klingele (1998) as follows: "At *aanaa* [smallest administrative unit] level 0.20/kg, Awadaye 5.60/kg (finance), Harar Region 0.20/kg, Jijiga 3.50 plus 0.20/kg, Togochale 0.50/kg, totaling over 10 birr per kilo of [khat] from the producer area to the Somali border" (p. 13). Most traders do not have to pay the entire gamut of these taxes unless they are taking the khat to the border with Somaliland with the intention of smuggling it. That possibility was in any case undermined by the resumption of legal export to Hargiessa in 1998. Traders paid varying levels of taxes depending on the number of tollgates they must clear before their destination. From Aweday, whether the traders headed west to Dire Dawa or east to Harer, each trader paid an average of 8.90 birr/kg in taxes. As the result, the amount of domestic taxes from the Harerge highlands climbed to an average of 22.7 million per annum between 1992 and 1997 and to an annual average of 203 million between 1999 and 2004 (Table 5.7).

Raising taxes did not ensure money flowing into the government's coffers. As Milkessa Wakjira (2004) points out, "the amount of khat that is taxed is not exceeding 13 percent of the estimated annual production of khat in eastern Ethiopia." At that level, the amount of tax collected in 2004 would be 175 million birr, a little below our estimate of the five year average of 203 million birr for the period 1999-2004 but consistent with other estimates (Anderson et al., 2007, p. 37). Nevertheless, it cannot be stated that the government had in place an effective tax system that could collect all the legal taxes. About 70 percent of potential tax income is lost to an ineffective system, rampant corruption at tollgates, and massive evasions at collection points (Wakjira, 2004).

TABLE 5.7 CHANGES IN KHAT INCOME TAX, 1965-2004

	Export Volume (Avg. MT)	Local Volume (90% in MT)	Total Production	Locally Marketed (60% in MT)	Income Tax Rate (Avg/kg in Birr)	Potential Revenue '000 Birr	Actual 30% '000 Birr
1965-74	1,535	13,815	15,350	9,210	1.00[a]	9,210	2,763.0
1975-91	2,336 (52.2%)	21,024	23,360	14,016	2.00[a]	28,032	8,409.6 (200.4%)
1992-98	3,159[c] (35.2%)	28,431	31,590	18,954	4.00[b]	75,816	22,744.8 (170.5%)
1999-04	12,665[c] (300.9%)	113,985	126,650	75,990	8.90[b]	676,311	202,893.3 (792.0%)

a. Based on Gebissa 2004. b. Based on Klingele (1998), Anderson et al., 2007, pp. 37, 56; Gashaw, 1999, p. 55. c. CSA, 1992-1999, 2000-2004.

Even with such inefficiencies in the system, the amount of state and municipal income tax revenue in eastern Harerge grew by nearly 800 percent to 202.9 million birr after taxes were raised in 1998 over the annual average of the previous five years. This amount is about a third of the 690.3 million khat tax revenue collected in 2004 nationwide. This is a substantial amount for state and local governments and indicative of the role of khat as a major source of funding for state government expenditure and development projects. Municipal governments also use khat revenue to finance minor infrastructure development activities, maintenance of public works, and to meet budget shortfalls. In 2003, for instance, khat revenue accounted for 60 percent of the total revenue for the city of Dire Dawa (Anderson et al., 2007, p. 37).

CONCLUSION

As the discussion in this chapter shows, it is impossible to ignore khat's micro- and macroeconomic benefits. Beyond the food security and prosperity that khat has enabled farmers to achieve, perhaps the most important aspects of the increasing cultivation of khat has been the ability of farmers to use the proceeds from khat sales to move to nonfarm occupations in the transportation, retail, and service sectors in urban centers. With ever increasing population growth and its consequent pressure on agricultural resources, it is difficult to envision that the available resources will continue to support nearly 90 percent of Ethiopia's population deriving its livelihood from agriculture. Under normal circumstances, farmers choose to remain connected to their ancestral land and continue the tradition of working the land. When they decide to abandon the land to which they are sentimentally attached, it is important to note that farm life has lost its future for each farmer who chooses to move to occupations in other sectors. Without any kind of extension support from the government, Harerge's farmers have shown that agricultural intensification and diversification are a great insurance against the inherent risks of subsistence farming for those who remain in the agricultural sector. By using the cash resources they obtained from khat sales to move to economic activities not connected to the smallholder agriculture based on meager and declining agricultural resources, they have shown what Ethiopia's Agricultural Development Led Industrialization (ADLI) strategy, a policy response to the country's food security and agricultural productivity challenge, should mean in practice.

NOTES

1. From a speech by Demissew Kassaye, Minister of Coffee and Tea Development (Addis Zemen, Jan, 15, 1983) and a popular song about coffee broadcast on Ethiopian radio every morning before announcement of coffee market prices the previous day. The lyrics go as follows.

The foundation of our economy
Coffee, coffee

> The source of our revenue
> Coffee, coffee
> The funder of our development
> Coffee, coffee
> The center pole of [our economy]
> Coffee, coffee

2. The survey figure is lower than the report from Eastern Harerge. Here the trend line is what is important than the accuracy of the figure. The figures of the local agricultural bureaus are the result of actual measurement and therefore more reliable than the national survey.

3. Habtemairiam Kassa et al. (2002) speak of three stages of change in farmers' response to market incentives and the consequent shift to cash cropping: feudalism prior to 1975, Marxism in 1975-1991, and free market economy since 1991. This typology follows political dispensation rather than change on the ground. Policies do influence decision-making on the farm, but changes driven by realities on the ground are more inexorable than policy-driven changes which could be reversed.

4. Based on reports from the Bureau of Agriculture from the Eastern Harerge and Western Harerge Zones of the Oromia Regional State, the Hareri Regional State, the Dire Dawa Special Administrative Zone, and the Somali Regional State.

5. At the time of my research in 1994, I met farmers who had sold 1 kg of khat for 1,100 birr that year.

6. The Ethiopian government has not been comfortable publishing accurate figures of khat production, consumption, or export earnings. Officials have occasionally blurted out some figures, which, in 2000, an independent newspaper was able to capture and report. "Khat has become the backbone of our country's economy" (Moresh, 2000, p. 1). The year 1999 given in the text refers to the Ethiopian fiscal year 1991, which begins on Septem-

ber 11, 1998, and ends on December 31, 1999, in the Gregorian calendar.

7. To my knowledge, there is a dearth of data when it comes to the amount of revenue generated by domestic transaction in eastern Ethiopia. There is no published data on khat tax and whatever haphazard information one can glean from sources at various levels of government is inconsistent. In this chapter, I have tried to use export figures, which are not reliable but consistent, to project the amount of domestic tax levels over a period.

Chapter 6

Agrarian Debacle and the Spread of the Dollar Leaf in Northern and Southern Ethiopia*

Degol Hailu

ಐಲ

Within the context of Ethiopia, the literature on khat mainly relates to the eastern regions of the country. However, khat cultivation, consumption and marketing have become omnipresent in the northern and southern regions of the country. Two of these are the Southern Nations Nationalities and Peoples Region (SNNPR) and the Amhara National Region (ANR) in the north. In the south, the main khat farming areas are the Sidama and Gurage Zones, while in the Amhara Region khat cultivation and consumption is becoming widespread in the districts surrounding the city of Bahir Dar. Khat from Wendo Genet in the south and Bahir Dar in the north, the *beleche* and *colombia* varieties respectively, are reputed for their high quality.

The literature on khat in these regions is meagre. One of the reasons for lack of research is the fact that these places are relatively new areas of khat trade and consumption. Government officials complain about the lack of a mandate to carry out studies. Academics express grief over the lack of funding for research. These shortcomings are the direct result of the controversy and official ambivalence surrounding the khat trade. While khat is castigated as a "narcotic" substance by national governments and international organizations, the benefits to farmers, traders and retailers are positively significant. The federal government of Ethiopia as well as the regional governments also profit from substantial tax revenues and foreign exchange the khat trade generates.

The history of the khat trade is succinctly presented in Gebissa (2004) and Anderson et al. (2007). In short, successive agricultural policies failed to make farming viable, beyond subsistence. Limited agrarian transformation has been catalytic to the growth of the khat trade. The recent spread of khat cultivation and consumption in Ethiopia is closely linked to the Agricultural Development-Led Industrialization (ADLI) strategy pursued for the last eighteen years.

This chapter argues that the ADLI has had limited success in resolving the low productivity features of agriculture in Ethiopia. While the strategy was successful in input provision and raising output levels, it has failed to lift farmers out of the subsistence quagmire. The strategy is not deemed successful from the perspective of the farmer. Agricultural inputs were costly, prices declined in boom years, and the terms of trade remained unfavorable. In this context, farmers switched to a high income cash crop, khat. A process of transformation which started in the eastern part of the country some 60 years ago, the adoption of khat as a cash crop, has spread even to the heartland of Christian Ethiopia where khat use, not so long ago, was seen as a Muslim habit. Khat presents farmers with an opportunity to improve their lives, as its prices are steadily rising; foreign and domestic demand is expanding; and the plant is relatively easy to grow and maintain. While governments at various levels do not support the expansion of khat plantations, they refrain from opposing it forcefully either. This is because the revenue that the khat trade generates supports a significant portion of their budgets.

THE AGRARIAN PARADOX

The discussion on khat cultivation revolves around substitution of coffee, whose prices have been fluctuating. For instance, Ethiopia's earnings from coffee dropped from 70 percent (US$4,330 million) of exports to 35 percent (US$165 million) between 1999 and 2004. In the same period, khat production doubled to US$58 million or 13 percent of GDP. The price of coffee fell by 17 percent from US$3 per kg to US$0.86. In contrast khat export prices average US$8 per kg.

The substitution of coffee with khat is explained by the dynamics of the political economy of agrarian intervention. As succinctly explained and dealt with in Ezekiel Gebissa's book – *Leaf of Allah*, first, the taxing of farmers and subsequent appropriation of land by successive settlers created the preconditions for agrarian transformation. Second, monopoly over agricultural marketing created another incentive for farmers to move into khat production. What is not explained in the literature is the continued stagnation of the agricultural sector in Ethiopia and how the agriculture-focused development strategy in the post 1991 period has not reversed the expansion of khat cultivation. For an extensive discussion on these issues (see Anderson, et al., 2007).

Ethiopia's agriculture-based development strategy, the ADLI has now been tried for nearly two decades. The strategy aims to increase land productivity through supply of vital inputs. These include extension services, fertilizers, subsidised credit, investment in rural roads, improved seeds and water management. Under the ADLI, it is estimated that more than one-third of the farming population benefits from inputs provided by the state. Extension services and fertilizer use have been rising steadily. From 1994 onwards, fertilizer use have increased significantly and reached 298,000 metric tons in 2000 from 107,000 metric tons in 1993. Pesticide consumption has also risen from 351 metric tons in 1995, to 436 in 1998, and 674 in 2001. However, irrigated agricultural areas have remained constant at 190,000 hectares between 1993 and 2002. For the same period, no change has been registered in the total tractors and harvester-threshers used in the farming sectors. Only 3,000 tractors and 100 harvester-threshers are in use (FAO, 2004).

In some years, notably 1996 and 2000, the strategy met some of its objectives. Farmers were able to increase annual yield of cereal and food crops. Although production of major crops and the area cultivated have increased, the stagnating yield figures and declining productivity paint a grim picture. Yield per hectare that averaged 1,170 kg in the 1980s stood at a lower 1,120 kg between 1990/1991 and 2003/2003. During the same periods, productivity per worker declined from an average 350 kg to 320 kg. Agricultural value added per worker fell from 310 to 266 birr.

The terms of trade have also been worsening for Ethiopia's rural sector. Prices for Urea, a type of fertilizer, have increased by 55.3 percent from 289.10 birr per 100 kg in 1993 to 448.90 birr in 2001, while the average price of cereals fell by 26 percent from 116.66 birr to 86.57 birr. The drop in the price of maize was the most notable, falling by 38 percent from 87 birr per 100 kg in 1993 to 54 birr in 2001. The most interesting story is the shift in relative prices rather than absolute fall in the prices of cereals. In 1993, for instance, 2.5 units of cereals were necessary to purchase 1 unit of fertilizer. By 2001 the situation had deteriorated, requiring about 5.2 units of cereals to obtain one unit of fertilizer. In 1993 3.3 units of maize were needed to purchase one unit of Urea. By 2001 more was needed – about 8.3 units of maize were required to purchase the same amount of fertilizer (Hailu, 2005).

The unfortunate outcomes of falling crop prices and shift in the terms of trade against farmers is explained by fixed demand for agricultural produce. Limited urbanization, underdeveloped agro-processing industry, and small export markets, have also curtailed the growth in demand. Paradoxically, productivity increases proved to be detrimental to farmers' income. Farmers could not even afford to pay the loan they have taken in the pre-boom years (Anderson et al., 2007).

The productivity paradox described above has not affected khat production. Acreage and yield for khat have been steadily rising, mainly related to the growth in domestic and export demand for the plant. While some years ago khat consumption was limited to certain religious and cultural groups, now we are witnessing cross cultural and transnational consumption. Khat is now pervasive among all ages, genders, social groups, income levels, and geographical boundaries. The export market has grown out of a consumer revolution in neighbouring countries (Djibouti and Somalia) and among the Ethiopian, Somali and Yemeni Diaspora in Europe and North America. The export market is now a complex operation with a range of intermediate activities including packaging, branding, retailing, and transport.

The author of this chapter interviewed a number of Agriculture Development Officers throughout Ethiopia, who have confirmed that no extension services such as credit, improved seed, and fertilizer are provided to khat. What captures attention is the fact that farmers are responding to the agrarian misfortune by diversifying into khat cultivation. This is because khat farmers seem not to suffer the same fate as cereal crops and coffee producers. Khat cultivation is attractive to farmers for various reasons. The plant is resistant to many crop diseases; it grows in marginal land and requires low labor inputs (Hailu, 2005). Farmers can also reap up to four harvests per year. Khat's net return per acre turns out to be greater than that from coffee. In Ethiopia, while khat accounts for only 13 percent of total cultivated land, it contributes 30–50 percent of farmers' total cash income per year (Hailu, 2007). It is also worth noting that khat cultivation is not covered under the ADLI, and there are no government policies designed to stimulate its expansion, but cultivation of the leaves has spread to the Southern and Amhara Regions, following in the footsteps of Oromia and Somali Regions in the eastern part of the country.

FIGURE 6.1: THE SOUTHERN NATIONS, NATIONALITIES AND PEOPLES STATE

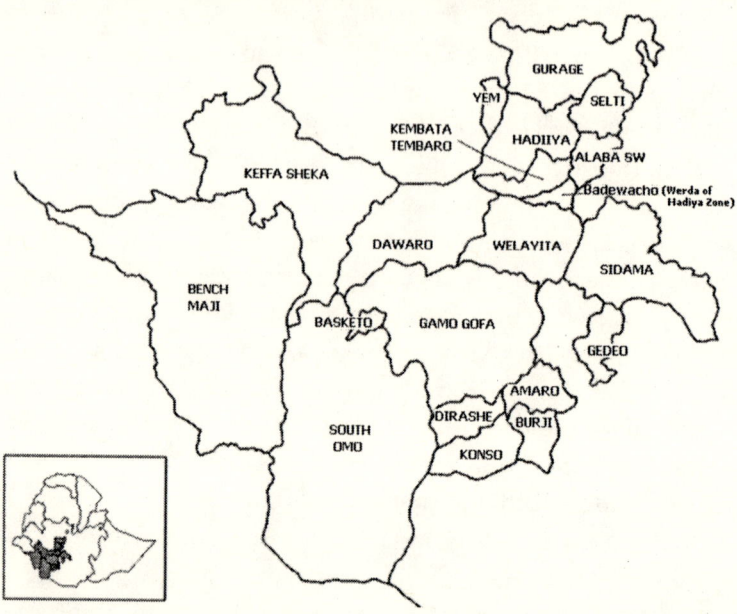

KHAT IN THE SOUTHERN REGIONAL STATE

The region's mainstay is agriculture with 23 percent cultivated out of the total area of 113,539 sq. km. About 20 percent of the land is earmarked for grazing, fairly distributed among the nine administrative zones and five districts. Some estimates show that 13 percent of the remaining land is cultivable. The region suffers from shrinking land size and poor asset ownership, with an average of 91 persons per square kilometre or agricultural density. The density ranges from 394 in Gedeo, 295 in Sidama, 258 in Hadiya, 197 in Gurage and 14 in Bench-Maji (Anderson, et al., 2007).

The economic data for the region is not favourable. About one million people were affected by draught between 2000 and 2004. Various household surveys show that a significant portion of the population lacks the productive assets necessary to lift them out of poverty. For instance, 72 percent of family heads own less than 2 hectares and 43 percent have no oxen to farm with. The food deficit of 8 billion kg has left the region dependent on food aid (BOPED 1998).

TABLE 6.1 KHAT PRODUCTION AND UTILIZATION IN ETHIOPIA, 2001/2002

Region	Total Production (000 kg)	Household Consumption %	Seed %	Sale %	Wages %	Animal Feed %	Others %
Tigray	11	88.7	--	88.7	--	--	0.7
Afar	--	--	--	--	--	--	--
Amhara	9,674	17.2	1.4	75.7	0.1	0.0	5.6
Oromia	451,890	38.5	0.6	54.4	0.4	0.1	6.0
Somali	96,836	27.9	0.6	64.4	0.2	0.0	6.9
Benishangul-Gumuz	1,089	41.8	2.5	52.4	--	--	3.3
Southern Region	206,712	18.0	0.4	78.5	0.3	0.0	2.8
Gambela	--	19.0	--	78.3	--	--	2.8
Hareri	21,337	29.7	1.2	61.3	0.0	0.0	7.4
Addis Ababa	--	23.0	--	77.0	--	--	--
Dire Dawa	--	49.5	0.8	39.4	0.8	0.0	9.5
Averages	787,549	35.3	1.0	67.0	0.3	0.0	5.0

Source: CACC (2003b)

The SNNPR is known for producing export commodities, mainly coffee and hides and skin. *Enset* (*Ensete ventricosum* or *E. ventricosum*) remains a staple food and is mainly destined for local consumption. Recent trends indicate that khat production is increasing rapidly, particularly in Sidama and Gurage Zones. The town of Wendo Genet is known for high quality khat such as *Anno, Beleche, Chenege Gulba, Nole and Sike*. These varieties are chewed by people as far as Mojo, Adama and Addis Ababa. They are also shipped to Kenya from market areas of *Aposto* and *Alata* on the Ethiopian side of the boarder. The khat is transported by trucks and beast of burden. Chewers consider the Gurage khat as good as Wendo Genet's, and it has found an expanding market in Addis Ababa.[1] An earlier study showed that about 12,672 hectares is devoted to khat cultivation and about 35 percent of households produce khat in Gurage Zone of SNNPR. In 2004, about 6,524,421 kg of khat was produced in this same zone on 5,620 hectares or 59 percent of the zone's cultivated land. The major khat producing areas are Enemor and Ener, Meher and Aklil, Kokir, Abeshege (Goro), Cheha, Endebir, Gunchere and Esia (PEDD 1998).

Nearly 8 million kg of khat was produced in 2001/2002 nationally (Table 6.1). The Oromia Region ranks first with production of 451 million kg followed by the SNNPR with over 206 million kg of khat production. In the Somali region production amounted to 96 million and 9 million kg in the Amhara Region. By far the largest khat consumption by the producer is registered in the Oromia Region with 38.5 percent used for local consumption. In Sidama, about 67 percent of the khat was produced for market.

Interviews with khat farmers in Sidama revealed that better living standards are associated with khat farmers. Farmers use khat proceeds to purchase consumer goods such as salt and kerosene lamp and improve housing using corrugated iron sheets. This is despite khat receiving no fertilizer, no credit, no seed and no input from the government's extension services.[2] While khat is predominantly produced in Gurage and Sidama Zones, other zones in SNNPR also cultivate it, but in small quantities.

Between 1999 and 2004 the total cultivated area and production covered under the government's national extension program were 2,042,211 hectares and 52,968,667 kg (Table 6.2). Of the areas covered with khat in the SNNPR, about 1,806 hectares fell under the extension program and 255,940 kg of khat was produced. However, as a percentage of the national area, only 0.0009 percent of the land covered by khat fell under the extension program. This means an insignificant portion of the area covered with khat cultivation received public attention. The bulk of the khat grows outside of the land assisted by the extension program.

TABLE 6.2 AREA AND PRODUCTION COVERED UNDER EXTENSION PROGRAM

	Total National Agricultural Land (ha) & Production (kg)		Khat		Khat: Southern Region % of total	
	Area	Production	Area	Production	Area	Production
1999/2000	1,875,146	47,574,366	1,781	252,560	0.0009	0.0053
2000/2001	2,039,864	49,556,260	1,828	254,680	0.0009	0.0051
2001/2002	2,068,464	51,672,523	1,798	256,241	0.0009	0.0050
2002/2003	2,088,485	53,758,967	1,806	257,510	0.0009	0.0048
2003/2004	2,139,097	62,281,219	1,815	258,710	0.0008	0.0042
Averages	2,042,211	52,968,667	1,806	255,940	0.0009	0.0049

Source: Anderson et al. (2007)

This finding is not surprising. One particular feature of the SNNPR khat industry is the tendency of the regional government to discourage its consumption and by implication its farming. This sharply contrasts with the Oromia Region where government officials are reluctant to speak out against khat consumption. In the SNNPR, government officials speak of future strategies to encourage farmers to switch to other crops other than khat. Various incentives are in place. Some of these include tax breaks for non-khat products, including poultry, vegetables, fruits and silk farming as well as livestock rearing. These incentives are accompanied by provision of quality seeds and breeds. The general thrust of public policy is to substitute perennial crops with annuals, also motivated by the belief that the latter provide better nutrition. Focus group discussions, moderated by the author of this chapter, found that in Gurage Zone some farmers have been able to benefit from these incentives and reduce their khat production. This is explained by the relatively low quality khat produced in the zone and the competition it faces from high quality khat from Harerge and Sidama.

THE KHAT DISTRIBUTION NETWORK

Most of the khat trade in Sidama Zone takes place at night. The markets are vibrant at *Teffera* and in *Basha* in Wendo Genet and *Tula* near Yergalem. *Teffera, Basha* and *Tula* are places that mesmerise the visitor. The early markets start at 7:00pm and some start late at 10:00pm with candle lights and torches. It is not an accident that the trade is conducted at night. Khat is a perishable plant and wilts relatively quickly. The night markets are meant to avoid sunlight and heat. The transactions last approximately six hours. Officials estimate that 800,000 birr worth of khat transactions take place on a daily basis in all three places. The practice is simple – the farmers line-up and wait for the buyers who come with their tracks from the various cities across the country, as far north as Addis Ababa, located 220km away. The khat trade does not stop at buying and selling. There are dealers, loaders, cutters, packers, including the restaurants, cafes, and bars. The khat trade has spurred a significant economic industry.

The distribution system in the Gurage Zone is characterised by the following: 1) some khat is directly retailed by the farmers on *ad hoc* basis – in the streets and by word of mouth; 2) the youth from the city centres buy from farmers and retail in urban centres; and 3) the farmers' siblings and close relatives dominate the distribution network up to Addis Ababa and supply to established shops. These groups are well organised and represented by two associations. The farmers and traders from the Gurage Zone are aided by continuous migration back and forth between the major city centres of the country and their home town. The Gurage traders are also known for their entrepreneurial acumen. This has helped consolidate the distribution network along kinship lines. In contrast, the market in SNNPR is characterised by the night time trading and formal retail-wholesale relations among the farmers, dealers and shop-owners.

TABLE 6.3 TAXES FROM KHAT AND TOTAL TAX REVENUE FOR 2001/2002, SNNPR

Zone	Khat Revenue	Total Tax Revenue	Khat as % of Total
Gurage	13,052,291	33,970,697	38.4
Sidama	8,674,209	35,794,993	24.2
Hadiya	371,621	13,992,337	2.7
Silti	168,052	7,744,526	2.2
Derashe	100,186	1,252,455	8.0
Gedeo	61,112	11,806,666	0.5
Amaro	9,214	1,257,543	0.7
Gamo Goffa	730	15,976,714	0.0
Bench Maji	830	7,826,517	0.0
Total	22,438,245	129,622,448	
Average			8.5

Source: Awassa Finance-Bureau (1996)

TABLE 6.4 KHAT TAX: VOLUME AND REVENUE, SNNPR

Year	Gurage (kg)	Sidama (kg)	Hadiya (kg)	Gedeo (kg)	Other Zones (kg)	Total Taxed Amount (kg)	Total Revenue (in birr)
1985/86	2,739,906	69,041	--	--	--	2,808,947	5,617,894
1986/87	4,162,904	129,398	99,553	--	--	4,391,855	8,783,710
1987/88	4,123,691	188,901	149,651	--	--	4,462,243	8,924,486
1988/89	5,730,710	237,864	87,831	--	--	6,056,405	12,112,810
1989/90	4,171,574	165,581	9,219	--	--	4,346,374	8,692,748
1990/91	5,056,806	180,521	74,445	--	--	5,311,772	10,623,544
1991/92	6,743,897	222,127	59,072	--	--	7,025,096	14,050,192
1992/93	5,593,999	341,615	165,910	27,615	193,525	6,101,524	12,203,048
1993/94	6,307,994	465,753	312,510	25,649	338,159	7,086,257	14,172,514
1994/95	9,000,957	1,181,869	286,311	32,022	318,333	10,469,137	20,938,274
1995/96	13,052,291	8,674,209	371,621	61,112	711,745	--	--
2001/02	6,526,146	4,337,105	185,811	30,556	--	11,079,617	33,238,851

Source: Awasa Finance Bureau (1996).

KHAT TAXES AND GOVERNMENT REVENUE

The SNNPR has a fiscal deficit of about 70 percent. In 1997, the regional government amended previous tax laws and increased the tax on khat from 2 birr per kg instituted in 1987 and replaced it with 3 birr per kg for domestic market and 5 birr for export.[3] As Table 6.3 shows 38 percent of the total tax revenue in Gurage Zone came from khat, followed by 24.2 percent in Sidama.

The introduction of value added tax (VAT) on coffee trade also contributed to a shift to khat production. As shown in Table 6.4, in 1985/1986, about 5.6 million birr was collected from taxing khat transactions. In 1990/1991 the amount increased to 10.6 million birr. By 2001/2002 the regional government earned 33.2 million birr from 11 million kg worth of khat taxed in the SNNPR. The highest tax revenue was obtained from Gurage Zone, which traded 6.5 million kg of khat in 2001/2002 followed by Sidama Zone at 4.3 million kg of khat. Tax evasion and smuggling are common and the government has introduced various measures. These included frequent transfers and rotation of customs offices; appointing secret agents; and frequent raids.

KHAT IN THE AMHARA REGIONAL STATE

With a population of 16.8 million, the Amhara Region covers a large area in the northern part of Ethiopian. The region is affected by recurrent droughts, sometimes leading to outright famine (ANRS-BOFED 2002). Khat consumption, long associated with the Muslim north-east and eastern parts of the country, has been a taboo in this predominantly Orthodox Christian region. What attracted attention recently is the spread both in the cultivation and in the consumption of khat in this region. A recent report (CACC 2003a) stated that "it is needless work to verify by quoting statistical evidence that coffee is a major foreign exchange earner. It may not be even surprising to hear that khat farming is becoming a rapidly expanding phenomenon in Ethio-

FIGURE 6.2: THE AMHARA REGIONAL STATE

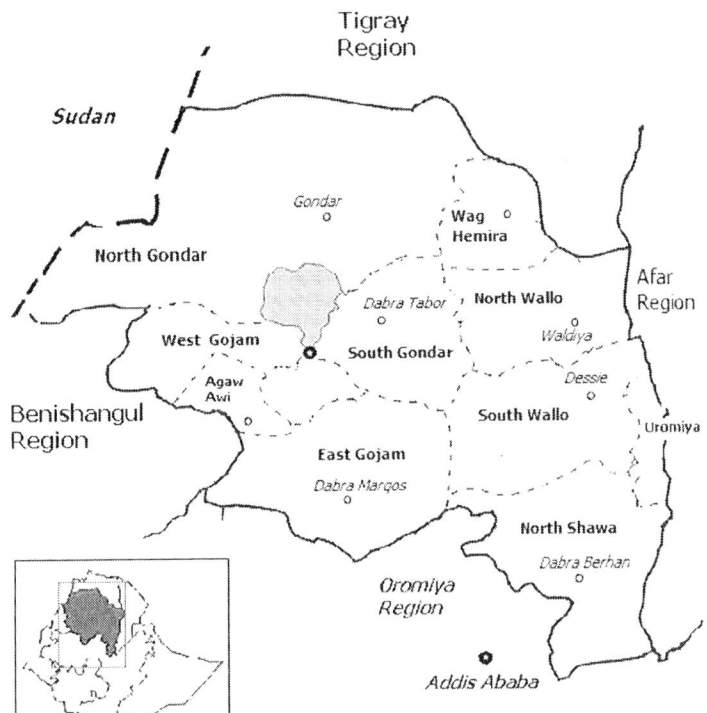

pia because of its economic importance..., the statistics on khat would perhaps be more tantalising than the others" (p. 6).

The story is familiar with what we described nationally and in the SNNPR region. Farmers in the Amhara Region are responding to long-term declines in agricultural productivity and constant threats of drought. Khat is relatively drought resistant and offers stable income.. Population pressure, land fragmentation, and declines in prices of other agricultural produce gave further impetus to substitutions by khat. These were confirmed by the Region's Finance and Economic Bureau officials during an interview with the author of this chapter. For instance, household size in the region is estimated at 5 members, while land holding is 0.5 hectares per household. It proves futile to increase productivity in such small land allotments.

TABLE 6.5 AREA AND PRODUCTION OF PERMANENT CROPS, AMHARA REGION, 2001/2002

	Crop Type	Area		Production		Yield
		HA	%	000 kg	%	000 kg
Amhara Region	Khat	2,008	11	9,674	13	4.8
	Coffee	6,264	34	12,772	17	2.0
	Hops	10,426	56	54,740	71	5.3
	Fruits	1,907	--	18,360	--	9.6
	Other Permanent	1,023	--	9,582	--	9.4
	All Crops	21,627	--	105,128	--	4.9
National	Khat	97,604	26	809,497	30	8.3
	Coffee	256,545	69	1,665,791	61	6.5
	Hops	19,098	5	253,384	9	13.3
	Fruits	36,780	--	2,041,517	--	55.5
	Other Permanent	13,007	--	990,790	--	76.2
	All Crops	423,045	--	--	--	--

Source: (CACC 2003a & CACC 2003b)

TABLE 6.6 AMHARA REGION: CROP PRODUCTION AND UTILIZATION, 2001/2002

Type of Crop	Total Production (000 kg)	Household Consumption %	Seed %	Sale %	Wages %	Animal Feed %	Others %
Cereals	33,762,051	68.4	12.8	13.0	2.5	0.4	2.9
Pulses	5,075,582	61.2	15.8	18.7	1.9	0.6	1.9
Oil Seeds	872,158	38.9	8.4	49.4	2.5	0.1	0.9
Vegetables	436,531	75.8	1.5	20.1	1.3	0.0	1.4
Root Crops	4,999,216	62.7	14.3	21.4	0.7	0.1	0.9
Permanent Crops	105,128	58.5	0.6	37.7	0.2	0.0	3.0
Fruits	18,353	39.8	0.7	73.9	0.2	0.4	1.8
Khat	9,674	17.2	1.4	75.7	0.1	0.0	5.6
Coffee	12,772	61.5	0.3	36.6	0.3	n/a	1.3

Source: (CACC 2003a)

As Table 6.5 shows, about 22 thousand hectares of land in the Amhara Region was covered with permanent crops. The largest proportion of this is taken by stimulant crops. About 56 percent of the area is devoted to hop production, followed by coffee 34 percent and khat 11 percent. Of the total stimulant crop production, 12.5 percent was khat, which remains lower than hop (70.9 percent) and coffee (16.5 percent). However, the yield for khat is 4,818 kg compared to 2,039 kg for coffee. The yield for hop stands at 5,250 kg. Other permanent crops grown in the region include fruits covering close to two thousand hectares, sugar cane one thousand hectares and *enset* ten hectares.

Table 6.6 shows production, consumption and sale of khat in Amhara Region. About 75.8 percent of the vegetables produced in Amhara Region were used for household consumption, followed by 68.4 percent for cereals, 62.7 for root crops, 61.5 for coffee and 17.2 percent for khat. On the other hand, 75.7 percent of the khat produced is used for sale compared to 36.6 percent for coffee. These figures confirm that khat consumption is new in the region and the consumption habit has not yet taken hold among the population. The bulk of the khat is produced for sale, indicating income-related incentives for its cultivation. Khat is cultivated throughout the Amhara region, mainly in Bure, Finoteselam, Dangla, Merawi, Mota, Meshenti, Hamusit, Adet, Delgi, Zenzelma, Kinbaba, Tekle Dengay, Zeghe, and Endasa.

In what follows, we closely look at khat production in Bahir Dar city. Out of the 17 administrative districts of Bahir Dar city, khat is cultivated in 13 of them. The remaining four are non-agricultural areas at the heart of the city and most of the retail trading is conducted in these four districts. Recently, they have shifted entirely to khat cultivation mainly due to coffee bearing disease, coupled with price fluctuations. Traditional crops such as corn, wheat, and *teff* are being replaced by khat. Khat seeds, *Colombia* and *Zemet*, are now marketable in the city centre.

In 1997 a survey revealed that, in Bahir Dar, the capital of the Amhara Region, each farmer dedicated 400 sq.m of land for khat on average, with a total of 62,950 sq.m of land covered by khat farming. The survey also ascertained that khat farmers have better housing, corrugated iron sheets as opposed to straw huts. The survey also found that about 20.3 percent of farmers rely on khat as their primary source of income (CBDSZ, 1997).

The above mentioned survey also identified about 1,259 khat farmers operating in Bahir Dar city alone. These farmers have used 157,067 cubic metre of water in 1997. Of these, two-thirds of the farmers were using pipe water, 13.9 percent utilised well water, and 19.06 percent directly fetched water from river Nile to water their precious shrubs.

Residents of Bahir Dar are in dispute about the use of scarce water. Non-khat farmers and nearby residents expressed their concerns about the shortages of drinking water, mainly blaming the diverting of water on khat farming. The dilemma is this: while the local economy is benefiting from the khat trade, the use of water for its cultivation is considered a "waste" by those not directly connected to the trade.

Perhaps, the following real case demonstrates the point. Kasanesh Biyadge, who was interviewed by the author of this chapter, is a 58 years old woman who planted khat eighteen years ago at the front of her house, within her compound in the suburb of Bahir Dar. Her customers are students of Bahir Dar University whose dormitory is half a mile away. She obtains about 240gm of khat every 20 days, which she sells at 3 birr, well below the market price, which is 6 birr for 100gm of khat. Kasanesh uses the money to finance her children's education. The children's school is 15km away and most of the income earned from khat sales is used to pay for one-way transport for the children. They walk whichever way they see fit.

However, things have recently changed for the worst. Bahir Dar University began expanding and a major construction got underway. The major road to the university happened to pass by Kasanesh's house. The distance from the road to her makeshift fence is less than two meters. This road is not asphalted and every passing truck from the construction blows massive dust.

Kasanesh's khat is sprayed with dust and turns from green to a brown shrub. The inevitable consequence is that the number of her customers dwindled; no one wanted to chew dusty leaves.

Kasanesh started to wash away the dust every night to woo back her customers. She succeeded and all was well, until the government decided to increase the price of water from 50 cents per cubic meters to 3 birr. For Kasanesh it was a lose-lose outcome. She is no longer able to afford the water bill incurred washing off the dust from her khat. She is giving up cultivating khat and contemplates shifting to growing vegetables and fruits. In the meantime her sons began the trip to school on foot both ways: 30 km each day of school year.

Kasanesh estimates about twenty households in the neighbourhood face the same predicament. They have repeatedly reported the case to local officials, but their pleas fell on deaf ears. Because the expansion of the university is likely to continue, Kasanesh and the other families simply have to switch to other activities. The sad part of all of this is the luck of consultation with the community before the construction begun.

As is the case in the rest of the country, the government is in a policy dilemma: it wants the revenue from khat but at the same time attempts to discourage its production by extending extension services to all farm produce, but khat. Extension services are given to farmers to switch back to annual crops. For instance, support for this endeavour is provided by international NGOs such as Swedish International Development Agency (SIDA), German Agency for Technical Cooperation (GTZ) and International Fund for Agricultural Development (IFAD). The initiatives included distribution of 10,000 mango seeds; passing of a new law to encourage fishing; training 260 experts and close to five thousand farmers in bee keeping; and in dairy farming and poultry. About 954 experts were trained in water use and in fruit seed improvements. The Amhara regional government also established 56 seed improvement stations; and training on pest control is frequently provided.[4] At the time of data collection for this chapter in 2004, these initiatives were in the process of implementation. Future research will, hopefully, reveal their impact on khat farming.

DISTRIBUTION AND CONSUMPTION

What stands out in the Bahir Dar khat market is the use of child labor. Children are involved in cutting, packing and retailing khat. The khat is often sold in the streets with roaming sellers. Some retail it among other consumables. There are also numerous khat vending shops. The expansion of the khat trade in the city is closely associated with the migration of the Gurage from the south and west of the country. As mentioned above, the Gurages are reputed for their networking prowess and entrepreneurial acumen. They not only retail khat in Bahir Dar, but also export it to Wallo, Gonder and Addis Ababa.

The CBDSZ (1997) survey identified 58 fixed shop retailers in Bahir Dar. Khat prices range from 3.50 birr to 7 birr per 100g, whereas high quality *Tana* and *Colombia* varieties sell for 10 birr. Khat chewers in Bahir Dar include the Muslim community, students, mainly from the Amhara State University, and civil servants. The federal civil servants and students, who mostly come from all corners of the country, are "blamed" by the conservative section of the Bahir Dar population for importing the chewing habit. The young chewers consume for entertainment purposes, while the unemployed and street children say they chew to forget their "misery". Focus group discussions also reveal that prostitutes and house wives chew and "pray" for a better life. The youth complain about lack of entertainment, sports facilities and youth clubs and cite these as reasons for chewing.[5]

TAXES AND GOVERNMENT REVENUE IN AMHARA REGION

The importance of khat in the Amhara Region is best illustrated by the revenue generated from its trade. As Table 6.7

shows, in Bahir Dar, revenue from khat has been steadily rising. In 1992/1993, about 70,359 birr was collected. By 1995/1996, the amount rose to 113,090 birr and by 2002/2003 the amount reached 128,208 birr. In the Amhara Region, tax tollgates were abolished in 1993 to allow free flow of goods.[6] At the time of research for this chapter, the regional government was in discussion to reintroduce the tollgates to collect taxes.[7]

TABLE 6.7 BAHIR DAR: TAX REVENUE FROM KHAT, BIRR*

Year	Revenue from Khat Tax
1992/93	70,359
1993/94	85,491
1994/95	102,177
1995/96	113,090
1996/97	113,980
1997/98	116,942
1998/99	120,511
1999/00	141,697
2000/01	126,346
2001/02	149,992
2002/03	128,125
2003/04[a]	128,208

Source: (BFED 2002). *July to January.

Currently khat tax is imposed on retailers. At the regional level, in 2003, 110,065,708 birr was collected from khat taxation, levying at rate of 2 birr for local consumption and 3 birr for export. This is equivalent to 8 percent of total tax revenues. By far, the largest revenue came from the South Wallo Zone of the Amhara Region, where about 175,226 birr was obtained. In the Oromia Zone in the same region, 11,953 birr was collected.

CONCLUSION

The trade in khat contributes significantly to farmers' livelihood in most parts of Ethiopia, including the Southern and Amhara Region. The national as well as the regional governments show tendencies denouncing khat consumption as a social evil. However, the benefit they generate from foreign exchange and tax revenues placed them in a paradoxical position, making them unable to develop a coherent policy.

One question to ask is why farmers cultivate khat. The answer lies in the agricultural policies pursued by successive governments. In this chapter, we have focused on the recent history. In the early 1990s, the Ethiopian government introduced a strategy known as ADLI. It was designed to increase land productivity through input provision. In some years and regions, the strategy succeeded in increasing use of fertilizers and pesticides and an almost 50 percent increase in cultivated area. Production of major crops nationally increased from 64 million kg before ADLI to 85 million kg afterwards.

The unfortunate consequence of the agricultural strategy has been to drive output prices down for major cereal crops. The country has been undergoing a slow process of urbanization. It has limited agro-processing capacity and weak export markets for its agricultural output. The terms of trade have moved against agriculture because input prices have grown faster than output prices. As a result, agricultural value added per worker declined in the non-khat sector from 310 birr in the 1980s to 266 birr during the period 1990/91-2002/03.

The Ethiopian economy is also propelled by coffee exports. However, earnings from it dropped from 2.1 billion birr in 1999 to 1.9 billion birr in 2004. Coffee prices declined from 135.23 US cents per lb in 1998 to 61.52 US cents per lb in 2002. Earnings from pulses and cereals, another export generating produce, have also declined while those from fruits and vegetables have remained low (see Anderson et al., 2007).

The upshot of the agrarian challenges described above has been to create incentives for farmers to switch their cultivation to khat. It is within this context that farmers in Southern and Amhara Regions increased the cultivation of khat. The plant is attractive to produce: it is resistant to many crop diseases, grows in marginal land, requires low labor inputs and can produce up to four harvests per year. Khat's net return per acre is much greater than that from coffee.

Farmers in Southern and Amhara Regions have benefited from the growth of consumption throughout Ethiopian and abroad. While traditionally confined to Muslim cultures, khat consumption now cuts across age, gender, religious, income and geographical boundaries. Mass consumerism is on the rise in neighbouring countries of Djibouti, Kenya and Somalia and as far away as Yemen and Uganda. Members of the Diaspora—Ethiopians, Somalis and Yemenis in Europe and North America—are also big khat chewers and source of foreign exchange earnings.

While the khat trade is vilified in many quarters, the evidence on its impacts on health is still inconclusive. The discussion on khat often bypasses the fact that its production, distribution and consumption are directly linked to a series of failures in agrarian transformation. A more constructive discussion is that which focuses on regulating its consumption through licensing retailers, setting age limits for consumption and establishing a system of quality control. The prohibition agenda threatens the livelihood of many farmers and traders as well as criminalizing them.

NOTES

* This chapter is based on research carried out for the book: David Anderson, Susan Beckerleg, Degol Hailu, & Axel Klein (2007) *The Khat Controversy: Stimulating the Debate on Drugs*, Berg Publishers.

1. There was a case whereby farmers in the Gurage Zone sprayed pesticide on their khat stock. This was to exterminate a worm

which has been eating into khat leaves. Sadly, this has caused outrage among consumers, who complained of stomach ache and a stronger than usaual euphoric effect.

2. Farmers in Silte, Gurage Zone, are currently planning to start irrigating their khat farms.

3. The tax processed khat is 4 birr per kg. One debated issue in the tax regime is concerned with packaging. The tax on khat is levied on total weight than net weight. The total weight includes packaging usually heavy water soaked grass wrappings. Effectively a portion of the tax is imposed on wrappings, which are of no use, but waste.

4. In this sector 29 NGOs have 62 projects in 67 districts participating in various extension support programs.

5. Interview with Ayachew Kebede: Bureau of Finance and Economic Development, Amhara National Regional State.

6. Interview with Assey Kebede: Revenue Bureau, Amhara National Regional State.

7. Interview with Akalu Geneme and Tilahun Eshete: Revenue Bureau, Amhara National Regional State.

Chapter 7

Khat and Livelihood Dynamics in the Harer Highlands of Ethiopia: Significance and Challenges

Habtemariam Kassa

෨෬

This chapter presents the major changes observed in the agricultural systems of the Harer highlands of Ethiopia, with special emphasis on the shift to a khat-based farm and regional economy. Data collected over the past two decades shows the agricultural system has shifted towards trees and shrubs, particularly khat, in the face of declining farm size and in response to growing market demand. This shift has promoted crop-livestock integration, improved farm income opportunities, expanded employment, and generated revenues for governments at various levels. These changes occurred without any external support, either from the government or donor agencies, primarily as the result of smallholder farmers' innovations based on intimate knowledge of the challenges and opportunities at the local level.

Yet policymakers and NGOs experts consider traditional mixed farming systems to be inherently unproductive, degrading to the environment, and incapable of evolving fast to meet the food demands of a growing population. They posit that, with declining availability of grazing lands, livestock grazing on marginal lands would have a negative impact on the environment and the viability of the farms themselves (de Wit et al., 1996). They also argue that increasing population pressure and low use of external inputs inevitably results in declining yields that would further aggravate an already precarious food security situation and eventually lead to the collapse of traditional food systems (Ruthenburg, 1980; FDRE, 1996; Quiñones et al., 1997). Consequently, introducing technologies that would increase crop and animal productivity faster than the rate of population growth are suggested to break the vicious cycle of demographic pressure and declining productivity (FDRE, 1996; PASDEP, 2007). This is a central theme of the national agricultural development strategy and explains the position of government policy makers (FDRE, 1996; PASDEP, 2007) and non-governmental development agencies (Quiñones et al., 1997).

Technology development and dissemination have accordingly favoured the use of high yielding grain varieties (e.g. hybrids) with fertilization and agrochemical control of weeds and pests and crossbred cows with improved management practices. Corresponding policies (e.g., credit schemes, importation of inputs) and extension programs are aimed at more grain-based farming systems as the focus of the Ethiopian Government has been on increasing food grain production (PASDEP, 2007). This strategy of intensification has been effective in increasing grain production in certain resource-rich areas of the country. In other areas, however, it has not been successful because the technological options were aggressively promoted across diverse agro-ecozones despite a host of biophysical, nutritional and economic interactions that constrain productivity and net economic returns (Nicholson et al., 1995).

Failure to understand farmers' circumstances, objectives, and strategies in the face of changing biophysical and socio-economic circumstances has been identified as a major factor in the failure of research agencies to generate technologies useful to the majority of smallholder farmers in Harer highlands (AUA, 1986; Wibaux, 1986; Mulatu & Kassa, 2001). Research has generally focused on increasing crop or livestock productivity through specialization while farmers have tended to intensify production through tree-crop-livestock integration and diversification of crop and livestock species. The result has been a dearth of appropriate technologies that farmers could embrace and deploy in their endeavor to adjust to the changing circumstances. It is with this observation that the study of the evolution of farming systems in the Harer highlands of Ethiopia was initiated in the early 1980s by the then Farming Systems Research Unit of Haramaya University.

The main objective of this chapter is to describe the evolution of the agricultural system in light of constraints and opportunities, and the role of khat in influencing livelihood dynamics in the Harer highlands. I argue that, despite the claims of the critics, the shift towards khat was unavoidable, given the absence of viable alternative crops. The importance of khat to household income, employment generation at regional level, and foreign currency earnings is highlighted to support my argument that khat production and marketing is likely to continue to be important to the livelihood systems of people in the Harer highlands as long as the demographic pressure on farm resources that precipitated the shift to khat remains in force. Yet I am acutely aware of the concerns recent data have raised in relation to the long term sustainability of an agricultural system dominated by khat. Agriculture in the Harer highlands is precarious. The imperative of future studies is to have a critical look into different investment options to help diversify economic activities to make people in the Harer highlands less susceptible to changes in the khat export market.

AGRICULTURE IN THE HARER HIGHLANDS: THE KHAT-LIVESTOCK INTERACTION

Southeastern Ethiopia is a region affected by increasing frequency of drought (FDRE, 1996), shrinking land holdings

(Adinew, 1991), and declining soil fertility (Hawando, 1982) that challenge the sustainability of smallholder farming. Farmers are known to adjust to changing conditions (Boserup, 1965). In this connection, Reijentjes et al. (1992) has pointed out that in their constant struggle to survive farmers gain a detailed understanding of the environment and develop site-specific and often very complex technologies to make optimal use of local resources. This leads to the development of wide ranging forms of agricultural systems adapted to their specific ecological and socioeconomic conditions (Holland, 1995). This capacity to trade off and adjust is one factor that determines the sustainability of agricultural systems (Conway, 1991), a concept that brings together mainly ecological, economic and social elements in development endeavors (DFID, 1999).

In the Harer highlands agriculture has been practiced for centuries (Westphal, 1975; Getahun, 1980) and farmers have made adjustments to their livelihood systems in the face of major changes in demographic, biophysical, and socio-economic environments (Mulatu & Kassa, 2001). During the last four decades, the size of the farming population has grown resulting in increased allocation of land for urban uses. The number of houses built in Kombolcha district of the Harer highlands increased by 30 percent between 1965 and 1983 alone (Langlais et al., 1984). Availability of land per capita (Adinew, 1991) and levels of soil fertility declined markedly (Hawando, 1982; Bojo and Cassells, 1995; Stork et al., 1997). During surveys conducted in 1985-87 and in mid 1990s, and in a discussion with key informants in 2009, farmers consistently identified in a descending order of importance the continuous shrinkage of land holdings, decline in soil fertility, unreliable rainfall pattern, and crop pests, and crop diseases as the major problems related to the farming subsystem.

In response to smaller land sizes and stagnant grain yields, farmers shifted to khat whose production and price increased markedly over time, and led to increases in household income (AUA, 1986; Gebissa, 1997; Mulatu & Kassa, 2001; Adinew, 2005). Cognizant of the biophysical limitations, economic and policy uncertainties, and the growing export market opportunities, farmers in the Harer highlands have historically subsisted on a multi-species portfolio of crops and livestock to produce many products and services. Crop-based mixed farming dominates. Major livestock species are cattle, sheep, goats and donkeys. Crops grown include sorghum (*Sorghum bicolor* (L.) Moench), maize (*Zea mays* L.), common bean (*Phaseolus vulgaris* L.), khat (*Catha edulis* (Vahl) Forsk. ex Endl.), sweet potato (*Ipomoea batatas* Lam.), potato (*Solanum tuberosum* L.), shallot (*Allium cepa* L. var. ascalonicum Baker), onion (*Allium cepa* L. var. cepa Helm). Other but less important crops are wheat (*Triticum aestivum* L.), barley (*Hordeum vulgare* L.) and vegetables such as carrot (*Daucus carota* L.), cabbage (*Brassica oleracea* L.), beetroot (*Beta vulgaris* L.), etc. Sorghum, maize, and common beans were the major food crops. They were intercropped in 90 percent of the cases and occupied the largest share of plots not planted with khat (Wibaux, 1986). The same were also planted between khat rows.

As khat harvested from plots fertilized by manure fetch higher prices, farmers' demands for manure grew over time (Kassa, Blake, & Nicholson, 2002). Given the upward demand for manure and greater opportunities for regional trade, the need for keeping livestock on-farm (own or leased) probably increased in recent times, although long-term farm level data are lacking to substantiate this. Livestock ownership should not be equated to the number of livestock physically present per farm in the Harer highlands, which has in fact declined as land per farm has decreased (Poschen, 1987). Declining farm size and expanded allocation of land for khat reduces feed supply for cattle but boosts demand for manure. This has led to widespread contractual agreements between farmers. Those who have the labor and feed but lack the capital to buy cattle instead lease from those who have more cattle than they could feed. This leasing of livestock helps both parties to match their available resources and generate more income.

As a result, in what is another kind of specialized system, khat and livestock in the Harer highlands remain important components of the rural livelihood system. But livestock roles

FIGURE 7.1 THE CROP-LIVESTOCK SYSTEM MODEL INDICATING THE ROLE OF KHAT IN THE SYSTEM

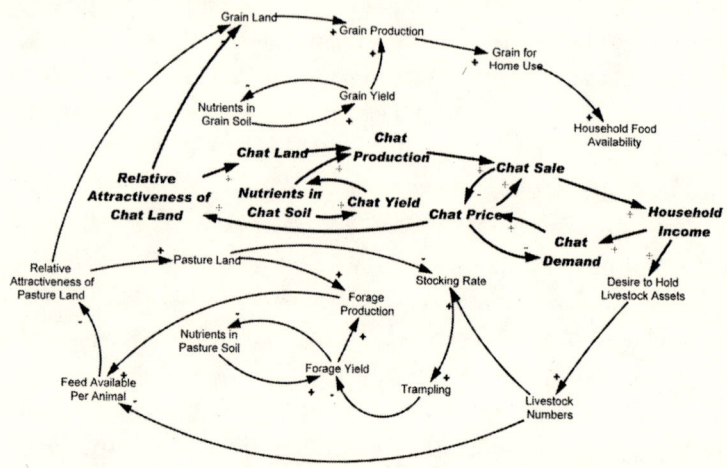

Source: Kassa, Blake & Nicholson, 2002.

are shifting from providers of food and draft power for plowing to providers of cash income (e.g., livestock fattened primarily for the export market), organic fertilizer (manure), and transport services. The implication of this for research is that, instead of maximized single output (e.g. milk) per animal, multiple livestock outputs would be more appealing to farmers. Any intervention to influence the dynamics of the livelihood system should consider crop-livestock integration (complementary and competitive relationship between grain land, khat land, and pasture land), the growing importance of khat and livestock and their influence on the relative attractiveness of land for different uses, the role of markets, and the growing importance of non-agricultural activities on household income (Figure 7.1). The crop-livestock model shown in Figure 7.1 depicts the relationships and feedback loops associated with khat framing. For instance, an upturn in demand for khat leads to a higher price, which in turn increases the attractiveness of land in khat. Besides,

FIGURE 7.2 ADAPTIVE EXPECTATIONS OF FARMERS WHILE MAKING PRODUCTION SYSTEMS CHOICES AND MANAGEMENT STRATEGIES

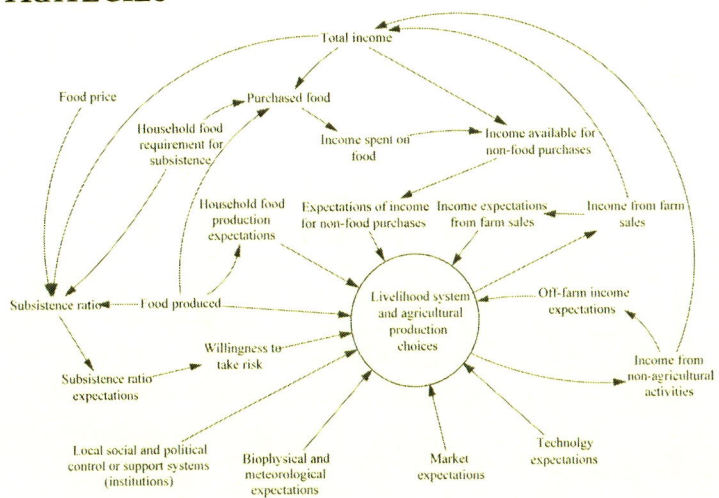

Source: Kassa, Blake & Nicholson, 2002.

improvement in household income increases the desire to have more livestock. A balancing loop also develops. The increase in the volume of khat sales reduces khat price. Another impact of the increase in the attractiveness of land for khat is a decrease in land used for grain production, and a reduction in the availability of food for the household. Understanding these relationships is a prerequisite to designing intervention options that will be acceptable to farmers (Kassa, Blake & Nicholson., 2002).

The smallholder farmers in Harer highlands have responded to the challenges they faced with a strategy of intensification, but in ways that are different from that of the policymakers and NGO experts. Thiers favors crop-livestock integration and diversification into non-agricultural activities. As noted above, such decisions are generally based on available resources and production objectives of farmers and their expectations of weather, production, markets or prices, and income (Figure 7.2). These factors that farmers take into account while making decisions are seldom fully understood by researchers and development planners. Inadequate understanding of resources at farm level and farmers' production objectives results in conflicts between mental models: what policy makers and development practitioners suggest and what farmers do, signaling misunderstandings about the crucial importance of interactions that govern farmers' behavior and changes in their livelihood strategies (Kassa, Blake & Nicholson, 2002).

Even though the long-term scenarios must be carefully studied, livelihoods in the Harer highlands have evolved in reasonably positive ways in terms of household income during the past three decades. More than any other agricultural commodities, khat production has given rise to a fast, efficient market and transport services with a daily flow of price and supply information between producers and the consumer and export market (Adinew, 2005).

ECONOMIC SIGNIFICANCE OF KHAT

HOUSEHOLD CASH INCOME

Given the small size of farm holdings and the current productivity levels, an average household could only produce between 0.6 and 0.8 tons of food grains, which is generally insufficient to feed an average family of about six. As a result, farmers purchase grains from nearby market centers; poor households purchasing proportionally more as compared to the amount produced on-farm (Kassa, 1986). Much of the money used to buy food items should have come from khat sales which constitute the major source of farm income (Kassa, 1986; Wibaux, 1986), indicating the role of khat in household food security. Though actual figures are lacking, tens of thousands of farming households are believed to have been directly engaged in the production and marketing of khat. Analysis of the comparative benefits of khat production and marketing at farm level provides insights as to why khat cultivation is expanding in the country.

Storck et al. (1991) reported that a third of sample households in the eastern Harer highlands reported to have earned a gross return between 1,000 birr and 5,000 birr per hectare during years of favorable weather. Wibaux (1986) reported that khat generates the highest return per hectare of cultivated land compared to any other crops grown in the Harer highlands, including coffee. In 1985-86, khat was the main source of cash income, accounting for 50 to 88 percent of the total annual cash income, depending on the wealth status of the household in the community (the wealth categories used being poor, medium and relatively wealthy), the lowest percentage being for poor households (Kassa, 1986). From a well managed and irrigated khat plot, in 1985-86 a farmer could obtain up to 14 times more, in terms of cash, than from growing sorghum or maize. With the 1985-86 annual average prices, the money obtained from the sale of khat from 1 hectare of land could buy about 15 tons of sorghum or maize grain (Mulatu & Kassa, 2001). By 1997-98, the khat-cereal value differential was even greater as the then price of "high quality" khat had quadrupled compared to the 1985-86 prices, while the price of cereals had increased by 20 percent at the most (Mulatu & Kassa, 2001). Adinew (2005) reported that about 70 percent of farmers' income in Habro district of Western Harerge in 2004 was obtained from khat. An average farmer in Kombolcha district of the Harer highlands earns more from khat in absolute terms but the proportion is likely to be similar to that of Habro as additional income is also earned from sale of vegetables which are exported to Somalia and Djibouti.

My rapid assessment in February 2009 shows signs of deterioration in the wellbeing of farming households, especially of those who were poor in the 1980s and 1990s. Interviews conducted with key informant farmers in Kombolcha district (same area where previous studies were conducted in mid 1980s and 1990s) shows that the trend in terms of the proportion of khat's contribution to total annual household cash income has remained the same but the amount of cash earned by a farming household has declined for those households whose khat plots have become increasingly fragmented in the process of dividing up and handing down to children. According to the reports of the key informants, some households who earned substantial income from khat in the 1980s and 1990s have experienced a slide into income poverty. It seems the impact of increased population pressure and subsequent reduction in farm size is being felt at least among households with large family sizes.

EMPLOYMENT CREATION AND LIVELIHOOD DIVERSIFICATION

Khat farming is expanding to the adjacent areas of major cities including Addis Ababa, Bahir Dar, Jimma, Shashemene and Awassa (Adinew, 2005; Dessie & Kinlund, 2008). As Degol Hailu shows in this volume, the price and market incentives are leading to expansion of khat even to areas where its cultivation has not been part of the traditional production system. These emerging marketing channels create new jobs for thousands of citizens involved in the collection, packing, transportation, and marketing activities. As Adinew (2005) observed, today one finds khat distributing and retailing shops in every city and major towns in Ethiopia.

Gebissa (1997) reported that since the 1970s, khat has been the fastest-growing and most profitable occupation involving millions of farmers, traders, and other service-providers in the whole Harer highlands. Gizaw (1999) reported that in 1997-98 in eastern Ethiopia alone official khat exporters employed 1,200 people in the packaging and transporting of khat. Many more get employed in khat markets in carrying, packing and brokering. The brokers get 10 percent of the price as a commission from sellers. Girls and women are engaged in khat retail trade in towns and major cities (Mulatu & Kassa, 2001). The gender dimension of this contribution needs to be recognized for an area where other employment opportunities for women are scarce. Gebissa (1997) observed that income earned from khat production and marketing was used to start non-farm occupations such as retail shops, taxi and pickup truck services, and, in some cases, restaurants, and hotels in the nearby towns. An excerpt from a report by Nita Bhalla of the BBC, based on the interview with a trader in Aweday, the major khat marketing center in the Harer highlands of Ethiopia, clearly illustrates the importance of khat production and marketing in creating remunerative employment.

> "I buy khat from the local farmers here in Awadai, and sell more than 200 kg every day. I send it to other areas in Ethiopia, like Addis Ababa and the Somali region and I also export it overseas. Khat is very good for Harerge. It is the backbone of the economy. If you compare khat farmers with other farmers, you will see their standard of living is so much better. They have good houses and some even have cars - how many farmers do you know that have cars?" (Nita Bhalla, BBC, 20 August 2002).

Discussions in 2009 with truck operators revealed that additional job opportunities are emerging for the youth in the construction sector as many people are building modern houses even in rural areas. All these indicate the role of khat production and marketing in employment creation in the Harer highlands and surrounding areas.

EXPORT EARNINGS AND TAX REVENUES

Khat makes significant contribution through foreign currency earnings. Earlier studies indicate that over 90 percent of production is sold (Wibaux, 1986), the rest being used for local consumption at household and village levels. Khat plots from which high quality khat is to be harvested will be managed differently. These plots normally receive attention in terms of input (water and manure) and care by men. Generally, men allocate certain portion of a khat plot for women to harvest and sell each time they need cash to cover routine household expenses. The main harvest of high quality khat is usually marketed by the head of the household, while khat sold for local consumption is usually handled by women. Klingele (1998) reported that in 1948 khat export totaled 200 tons, and in 1957 it reached 1,400 tons. During 1999 to 2004, an average of 11,000 tons of khat had been exported annually (Adinew, 2005). Even though accurate statistics on the trend in the volume and value of khat produced and marketing over years are lacking, khat has become an important foreign currency earner. Adinew (2005) estimated the contribution of khat to foreign exchange earnings. In southeastern Ethiopia alone, the contribution of khat to foreign exchange earnings has been significant. For the period 1989/90 to 1999/00 Ethiopian Fiscal Year, the annual average contribution of khat in relation to other commodities stood first (46.2 percent), followed by coffee (45 percent) and other commodities (8.8 percent).

The producer price for export quality khat is also known to have ranged from 20 to 50 birr per kilo, with an average price of around 30 birr in the 1990s (Klingele, 1998). During dry seasons where khat supply is scarce, the price increases to a hundred or more birr in places like Dire Dawa (Adinew, 2005). The key informant interviews in 2009 revealed that dry season prices have significantly increased. But such price increases are not reflected in the export figures as such prices are set by the Ethiopian Government in consultation with the Government of Djibouti. In 2005, for instance, the FOB price was set at $6 per kg. The export companies have complained that the FOB price in the dry season falls below the market price locally and does not cover costs of purchase and transport from the southern and western parts of the country. During the 1970s and 1980s, one government-affiliated exporting company had complete monopoly on khat exports, dispatching on average 8,000 kg per day to Djibouti. In Djibouti there has been only one sole importing agency. Currently, the largest exporter is the Dinsho Khat Export Company. Currently the government licenses only a limited number of companies for khat export, and each of the companies exports up to 5,000 kg per day.

Khat export from prominent production and marketing regions like Harerge keeps on growing. Adinew (2005) reviewed a 20-year data and concluded that export value of khat has increased over time. The export value which was 16 million birr in 1984/85, increased to 200 million birr in 1996/97, and reached 760 million birr in the year 2003/2004. The *Ethiopian Herald*, a government newspaper, reported that khat accounted for 10.5 percent of the $ 793.2 million total foreign currency earnings during the 2004/2005 Ethiopian fiscal year. The rise in export value of khat is attributed both to an increase in volume and a significant rise in export price that for example rose from 20,000 birr/tons in 1992/93 to 80,000 birr/tons in 2002/2003 (Adinew, 2005).

In recent years, khat export from Ethiopia has found new markets beyond the neighboring countries in Europe and Asia. But much of khat export to Europe does not go directly from Ethiopia but from other re-exporting countries. A study by the Home Office (UK) reported that in the first 6 months of 2005, approximately 5-7 tons from Kenya, 500 kg from Ethiopia and 175 kg from Yemen were arriving each day in the UK to be distributed by an efficient network to the khat using communities across the land. Khat reaches UK cities daily via European airports (mainly England and The Netherlands) as the trade in khat is a legal business. The UK has become a major destination for imports and exports of khat. Significant price differentials exist between UK and other countries, primarily the USA. In 2005, the average price in UK was about £15/kg of khat while in the USA, where it is illegal but marketed to Somali, Yemen, and Ethiopian communities, it cost $400/kg (Rawlins, 2005). These markets are likely to promote the production and marketing of khat in major producing countries like Ethiopia.

Khat is also a major source of income for the regional governments in Ethiopia. The federal government collects taxes from exporting companies that are required to record volume and value of sales (Adinew, 2005). Throughout the market chain when khat is moved from the point of production to the export market centers, khat is taxed by different local and regional authorities. In this regard Klingele (1998) reported that a khat trader, exporting to Hargiesa town of Somaliland, is taxed at district level by the local council where he purchased the khat (about 0.20/kg), at Aweday main khat market by the Central Government (5.60/kg), at Harer by the Regional Government (0.20/kg), at Jijiga (3.70/kg) by another Regional Government, and at Togouchale (0.50/kg when crossing the boundary), totaling over 10 birr per kilo of khat. Traders exporting to Djibouti via Dire Dawa are taxed at Dire Dawa as well. Recently however taxation procedures on khat have been simplified. Though the frequency of taxation and the tax level may vary from place to place, a substantial amount of tax revenue is generated by Regional Governments and City Administrations. As Klingele (1998) and Adinew (2005) remarked, no other agricultural product fetches as high a tax income as khat.

CHALLENGES OF LONG TERM SUSTAINABILITY

SOCIAL AND HEALTH RELATED CONCERNS

The rapid expansion of khat consumption culture in the country is raising serious social and health related concerns (e.g. Mulatu & Kassa, 2001; Adinew, 2005). These relate primarily to khat dependence and the time and money spent on chewing khat daily by a significant number of the country's population, particularly the youth. In discussing about the future of khat production and marketing, these concerns must be highlighted against its economic benefits.

EXPANDING CONSUMPTION AND LACK OF ALTERNATIVE INVESTMENT

In the 1980s, local stores brought sorghum and wheat grains, soft drinks, radios, and lumps for consumption by farmers. In the 1990s, bed sheets and blankets, mattresses, television sets and packed foods appeared. In 2009, poles and construction materials, charcoal, etc were being widely traded. Shops were stocking more and more of wheat flours, powder milk, rice, oils, detergents, and packed foods (spaghetti, macaroni, etc) and drinks. Vehicles purchased (by farmers) were no longer Land Rover trucks for rural transport and taxi services, but mini-buses and small dump trucks used to transport construction materials. Though in-depth studies are lacking, it is obvious to imagine that the consumption pattern is changing and possibly expanding. Evidence of investment to diversify economic activities outside khat marketing is lacking to open up alternative employment opportunities for many in rural and urban areas of the Harer highlands.

Given that khat is internationally considered a "mild drug", it is likely that importing countries may ban its import. If this happens, there is little chance that the domestic market would be as attractive as the export market. Adinew (2005) remarks that high benefits derived from khat might have somehow undermined diversification innovation and the resilience capacity of khat producing communities. This in turn slows down the development of sustainable, more productive and socially acceptable agricultural production systems that can overcome the challenges of resource scarcity, land/environmental degradation, poverty, and food insecurity. Unless farmers and major khat producing area administers start to diversify, the household, regional and national economy will be greatly damaged by any move to reduce or limit khat import.

SHRINKING FARM SIZES MAKING LARGER FAMILIES SLIDE INTO INCOME POVERTY

Families in the Harer highlands sub-divide their own plots each time their sons get married, as generally female heirs do not inherit land. This leads to further fragmentation of land to a point where enough cash income may not be obtained even if the price of khat continues to rise. This was observed during my 2009 visit to one of the well known producing area in Kombolcha district. Some farmers who were considered relatively well off in the 1980s and 1990s are now reportedly struggling to earn enough cash income to feed their children. This is a troubling development that needs to be studied further, but the trajectory is clear that the khat-based diversification strategy has reached a critical turning point.

MANAGING THE NUTRIENT BALANCE OF KHAT PRODUCING FARMS

As leaves and twigs are harvested and marketed, nutrients are transported to the nearby towns. Though the extent of this loss needs to be assessed, its long term effect is likely to be negative, especially if nutrient losses are not compensated for through organic matter application. As discussed earlier, one of the major objectives of framers raising livestock is to get manure for maintaining soil fertility. With livestock holding per household declining over time, nutrients that were transported away cannot be replenished. Some relatively wealthy households are known to buy and transport decomposed organic matter from nearby towns and cities. This calls for research exploring options that will lead to win-win scenarios for municipalities to remove urban wastes composed mainly of khat leftovers and khat producing farmers who really need these on their farms to compensate nutrient losses due to khat sales.

CONCLUSIONS

In response to changing biophysical, demographic and socio-economic environment, farmers in the Harer highlands shifted towards khat production. As Mulatu & Kassa (2001) concluded in the absence of feasible economic alternatives and given demographic pressure and resource limitations, the current evolutionary pathway of the farming system towards a khat-dominated farm economy was unavoidable. Despite its reported negative effects on health and other social aspects, khat has played important economic and possibly ecological roles for farmers in the Harer highlands of Ethiopia. Thus, the importance of khat production and marketing in increasing cash income and in improving household food security of the poor cannot be over emphasized. Khat has been and will continue to play an important role in the livelihood dynamics of people in the Harer highlands for the foreseeable future.

According to Wibaux (1986), Klingele (1998), Mulatu & Kassa (2001) and Adinew (2005), the major factors that contributed to khat-based agricultural economy in the Harer highlands included growing local and export markets, fast and efficient marketing and transport services, shrinking farm size, declining soil fertility, adaptation of khat to the ecological conditions (the hardy characteristics of the plant to withstand drought and frost, and disease and insect problems), and the low labor and input requirements for its production. The return per hectare is the highest of all crops.

The overall socioeconomic aspects of the production, marketing, and consumption of khat must be studied and presented in a systematic way to inform policy makers and development planners. As the system is evolving towards a more complex system, understanding multiple interactions among key elements of the livelihood system is necessary to come up with acceptable intervention strategies. Unless these are factored in the planning process, research, extension and development efforts will continue to mismatch with farmers' expectations and national interests. For countries like Ethiopia, as Kydd (2002) remarks, the dynamics of smallholder agriculture ought to be a central question for research and debates about development.

Addressing the challenge of agriculture in Harer highlands requires studying the long-term trends in the evolution of agricultural and livelihood systems, especially in terms of their limits and potentials for enhancing human welfare. Interactions among the major sub-systems must be well understood. Better understanding of the factors governing long-term behavior of livelihood systems, has both scientific and practical importance in poverty alleviation (Vosti, 1995). It facilitates making informed decisions about the intensification of production systems in ecologically and economically viable ways (Saeed, 1994; de Haan et al., 1997; Sterman, 2000; Nicholson et al., 2001).

As the khat-based household and regional economy is dependent on the export market, any import ban will have a devastating impact on the livelihoods of many in the Harer highlands. Its impact on the national economy will also be significant. Thus comprehensive studies are needed to understand consumption patterns and to identify options where the cash earned from the production and marketing of khat can be invested and the regional economy diversified.

Chapter 8

Market Incentives, Rural Livelihoods, and a Policy Dilemma: Expansion of Khat Production in Eastern Ethiopia

Tesfaye Lemma Tefera and Daniel Start[1]

෨෬

Khat consumption and trade which remained confined to Ethiopia, Yemen, Kenya, Somalia and Djibouti for centuries have become a global activity since early 1990s (Carrier, 2005; Gebissa, 2004; Kennedy, 1997a; Elmi et al., 1987). Advances in transportation and increased migration, particularly of Somalis, from East Africa to Europe and North America, were key developments that made the expansion possible (Anderson et al., 2007). The host countries of the new immigrants have responded in various ways to the khat phenomenon. The United States, Canada and Scandinavians have outlawed khat trade and use, citing the presence of the alkaloid cathinone, which has stimulating effects similar to that of amphetamine, in the leaves (Brooks, 1997). On the other hand, the UK government has decided to keep khat a licit substance based on the recommendation of the Advisory Council on the Misuses of Drugs (ACMD, 2005).

In Ethiopia, the cultivation of khat as a cash crop is expanding in the East Harerge Zone of the Oromia Regional State in response to rising demand for the leaf from the long standing and new export markets in Djibouti and Somalia as well as an expanding domestic consumer base. Farmers are increasingly shifting their meager resources to khat production spurred primarily by market incentives, but also in response to specific farming challenges, such as drought, production risks, and diminishing farm size, and quality against increasing subsistence requirement.

The impact of the shift made at the household level has had implications for the local, regional, and national economies. Earlier investigation in Haramaya (East Harerge), Kuni and Sabale (West Harerge) districts of East Harerge Zone (Tefera et al., 2003) has shown that khat crop cover ranges from 24 percent to 54 percent of total cropped land area, generates about 29 percent of the average annual income of farm households, and accounts for 85 percent of their total cash crop income. In the Habro district of the adjacent West Harerge Zone, khat covers 55 percent of the cultivated land and accounts for 70 percent of households' income (Feyissa & Aune, 2003). Khat export revenue as a share of GDP in Ethiopia averaged 1.7 percent for the 1990s while public health expenditure as a share of GDP averaged 1.2 percent (Anderson et al., 2007). According to the National Bank of Ethiopia (2005, cited in Ethiopian Economic Association, 2007), in 2004/05, the country earned $100 million in hard currency by exporting 22,000 tons of khat.

On the other hand, the rapidly growing culture of khat consumption in the capital, Addis Ababa, and other major urban centers, particularly among Ethiopian youth, coupled with widespread perception regarding negative social and health consequences of regular khat consumption has given way to public debates and a serious policy concern. Secondly, khat production competes with food and export crops like coffee for scarce farm resources, particularly land and water. Farmers need to buy cereal to meet their food deficits as land and other farm resource are increasingly committed to the production of khat (Feyissa & Aune, 2003; Tefera et al., 2003). This is an equally worrisome policy issue as the expansion of khat production might compromise food security of rural households and foreign exchange earnings from coffee.

Despite heightened national and global debates that the increasing khat production, trade and use have triggered, the evidence on its adverse effects is scanty (Beckerleg, 2008) and inconclusive (Al-Hebshi & Skaug, 2005). While the negative side effects of khat cannot be ignored, it must be emphasized that what is at stake in the debate is the country's export earnings and the livelihood of millions of poor households and individuals who earn their living directly or indirectly from the khat supply chain. These critical issues have often been overlooked. The purpose of this chapter is to inform the khat discourse from a rural livelihood perspective by providing empirical evidence on the factors behind farmers' decision to increasingly shift scarce farm resource to khat production. In addition, the chapter highlights the effects that the shift to khat production has had on income, food security, and poverty at the local level. The empirical evidence, which we believe is critical to informing a policy debate on khat, is based on data collected during a fieldwork conducted in March and April, 2005 as part of rural livelihood study in the Deder and Meta districts of the Eastern Harerge (Start et al., 2005).

THE STUDY DISTRICTS AND INQUIRY METHODS

The districts of Deder and Meta are among the seven districts in East Harerge Zone of the Oromia Regional State of Ethiopia. These districts are predominantly rural with a population density of over 300 persons per sq. km, which is far higher than

FIGURE 8.1 LOCATION OF EAST HARERGE ZONE

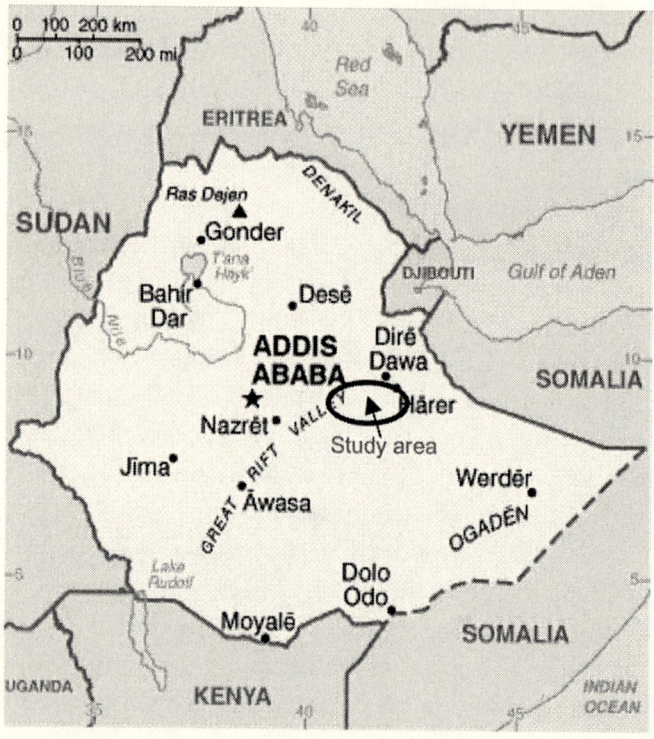

the average population density for the East Harerge Zone (about 100 per square kilometer). Deder and Meta districts are known to be drought-prone and have been major aid recipients over the past two decades.

Agro-climatically, the two districts are divided into highland (33 percent), midland (44 percent) and lowland (23 percent) with altitudes of 2300-3300, 1500-2300 and 500- 1500 meters above sea level respectively. Average annual rainfall in the two districts ranges from 650 mm in the lowlands to about 1000 mm in the highland areas. Traditionally the districts have had good quality land, relatively good irrigation, and are located close to a major motor road. A major highway links the region with Addis

Ababa (450 kms west) and Dire Dawa (100kms east). The latter is the second largest city of the country and an outlet to export markets with air and railway links, particularly to Djibouti and Somalia. The Harerge highlands, encompassing Deder and Meta, are characterized by rough topography and exposure to severe soil erosion. The farming systems are predominantly mixed crop-livestock systems. The major crops in the study area are maize, sorghum, khat and coffee. Approximately two-thirds of the cropped land area is sown to food crops, but at least two-thirds of the overall farm households' incomes are from cash crop.

The differences between highland, midland and lowland cropping systems are minor, with the big four crops, maize, sorghum, khat and coffee heavily cultivated in all areas. In addition, there are some sweet potatoes, fruit, and groundnuts in the lowlands, Irish potato and more coffee in the midlands, and more sorghum and khat but no coffee in the highlands.

The study was conducted in twelve *kebeles*[2]: two highland *kebeles*; two midland *kebeles*; and two lowland *kebeles* in each zone, involving 600 households. A range of criteria was used to stratify the *kebeles* in the two districts into six zones and then select sample *wirtus*,[3] households, and farmer groups for focused group discussions. One *kebele* was selected randomly from each of the six zones in each of the two districts. As the populations of Deder and Meta districts were close to equal, and the distribution of *kebeles* between agro-ecological zones was also roughly similar, we felt this would give a useful representative collection of 12 case study *kebeles*. From the 12 *kebeles*, six *kebeles* were chosen for case study analysis ensuring that low, mid and highland agro-ecological zones (AEZ) were covered, and a mixture of high and low level market access from each district. For the household sampling, 50 households were randomly selected from each sampled *wirtu's* updated households list. The analysis[4] reported in the next section presents primary empirical evidence on the factors that contribute to the expansion of khat, its roles in the livelihoods of rural households, and growth multiplier effects of incomes from khat in rural local economy.

EXPANSION OF KHAT

BROAD CROPPING PATTERNS

The steep topography and altitudinal range of the Deder and Meta districts provide for diverse cropping systems. Within one village crops can range from rain-fed barley and wheat on the upper reaches of the village hills, down through rain-fed maize and sorghum on the more gentle slopes, to the irrigated khat and coffee on the best land in the valley bottoms.

Sorghum is a traditional semi-arid crop and, being slightly more disease-resistant and drought-tolerant than maize, it is more suited to the warmer, drier lowland areas. However, both crops are popular and command similar amounts of yield and price. Both are frequently intercropped with horse beans.

Coffee (the major foreign exchange earner for the country) and khat are both cash crops with good returns. Coffee is more adapted to the lowland areas of the hills. Khat, on the other hand, requiring very little management, no inorganic fertilizer, and commanding very good price, is fast growing in popularity wherever there is ready access to markets and good irrigation. Khat trees are sometimes intercropped with maize when they are young.

Sweet potato is a common food crop in the lowlands. Traditionally it is grown as food for use in drought seasons, and dug up when needed, otherwise remaining intact in the ground. In contrast, Irish potato is a new crop with high market value. It has been taken up by the better-off farmers in one village soon after it was introduced by an NGO (Oxfam- Great Britain) eight years before the survey. Like many other horticultural crops, Irish potato needs good and consistent irrigation supply, high management inputs, and good market access, as transport is relatively difficult and bruising common.

TABLE 8.1 CROPPED AREA SHARES FOR MAIN FOOD CROPS *(PERCENTAGES OF AEZ/WEALTH CATEGORIES)*

	Maize	Sorghum	Total* Food Crops	Khat	Coffee	Total* Cash Crops	Total food and cash crops
Low	41	24	66	26	6	34	100
Mid	42	20	63	21	14	37	100
High	38	35	73	26	0	27	100
Impoverished	40	26	-	-	11	34	100
Poor	41	25	-	-	7	33	100
Better	42	24	-	-	4	32	100
Rich	42	21	66	25	5	32	100
Total	41	25	67	24	7	33	100

* Total food/cash crops include the other minor food/cash crops

Barley, wheat, groundnuts, sesame, tree crops, fruits and fodder crops also feature in Deder and Meta, but as indicated in Table 8.1, the majority of the cropped area is covered by the four main crops: maize (41 percent of cropped area), sorghum (25 percent), khat (24 percent) and coffee (7 percent). Thus khat is more than three times as popular as coffee, while maize is almost twice as popular as sorghum. Exactly a third of the total area is planted to cash crops, while two-thirds is planted to food crops. This split is surprisingly similar across all wealth groups. Only in the highland do we see more food crops being grown though these are often sold for cash.

There are further distinctions within the cash/food categories. Highlanders grow no coffee and almost only khat, while midlanders grow more coffee and less khat. This is because khat thrives in cooler and wetter highland climates and coffee does well in warmer, drier mid to lowland climates. Less significant cash crops, such as wheat, Irish potato, and groundnuts make up only 2 percent in additional cropped area.

LAND USE TREND AND CROP CHOICES

The average land holding size of households in the two districts is very small (one-third of a hectare) compared to the national average of about 1 ha, reflecting the high population density in the area. The trend in land use over the three years prior to the survey shows almost all households have increased their khat area at the expense of food and other cash crops, on average a 91 percent increase in khat area. By far the bulk of this increase has occurred in the highland areas where khat areas have more than quadrupled. The increases have applied to households of all wealth groups; and 89 percent of this new area has displaced maize or sorghum.

Khat production is particularly well suited to the research area for several reasons. Some of these are summarized in Table 8.2 and set against those of other crop choices. From farmers' perspective, the most important constraining factor of khat production after land is the lack of water for irrigation. Availability of irrigation has the potential to lower risk substantially, create higher yields and, most importantly, produce crops in dry season when prices are at their highest. A second consideration is input management, not only cost but also time and drudgery. Third, there is income, including market value or domestic value, if not marketed. Just as important as the total amount of income in the year is the cash flow and the degree to which it can be flexibly controlled to increase production in times of need and reduce it in times of plenty. Fourthly, khat is the most preferred and economically attractive conservation crop in the Harerge highlands (Tefera, 2003). According to Klingele (1998), khat is the only agro-forestry plant that could stabilize erosion control structures while yielding economic benefit, an important feature of sustinability. Moreover, as a crop that could be harvested for several decades with little input, khat is an important means of accumulation and intergenerational asset transfer, supplementing or substituting the declining cattle population due to feed scarcity. As a perennial crop, khat is increasingly seen as old age insurance as the prevailing tenure system does not permit the accumulation of land asset (Tefera, 2003).

Table 8.2 clearly shows how high the levels of risk are for almost all the cropping options. There are the risks from markets, drought, and diseases and pests. East Harerge farmers live in a high risk economic environment which makes investment difficult and irrigation a premium.

TABLE 8.2 CROP CHOICES AND TRADE-OFFS MATRIX

	Irrigation requirement	Management	Annual Income (birr[5]/quindi)	Market / Production risk
Khat	Medium Drought tolerant but irrigation best	Low Low labor, some manure compost	High 1500-6000 All year round profits, flexible harvesting.	High Variable seasonal prices. Perishable. Frost sensitive, but rarely affected by diseases.
Coffee	High Needs irrigation.	High Intensive Management requirement. Expensive Pesticides	Med-High 1500-5000 But good harvest only every two years.	High Coffee Berry Disease can devastate crop, stores well but variable price year to year.
Irish potato	High requirement	High Weeding, fertilization	Very high 2000-8000	Medium-high Market prices rising, but can be bruised
Maize	Low-medium Drought-sensitive but rarely irrigated.	Medium 25 kg urea and DAP/*quindi*, improved seed.	Low-medium 125-1000 But is also preferred food,	High Low drought tolerance. No need to market.
Sorghum	Low Drought tolerant and rarely irrigated	Medium No fertilizer	Low 100-800 Food crop and fodder crop	Medium Drought tolerant. Harbors diseases, stalk borer. Often not for sale.

Quindi (an area of land ploughed by a pair of oxen/ day/ in about six hours) is a local unit for measuring land. One hectare ~ eight quindi.

In the study sites, the shift away from coffee had been a more gradual process over the ten years prior to the survey. Coffee had been the cash crop of choice for several decades through the late 1990s, but falling world prices, the devastation of the Coffee Berry Disease (CBD), and the removal of subsidy for protection chemicals since the early 1990s have led to many farmers cutting down established coffee trees in favor of khat which enjoyed better prices (Tefera, 2003).

This situation is less severe in the lowland areas where poor market access makes khat marketing extremely difficult, if not impossible. In 2005, coffee prices were improving and the best farmers estimated that they could harvest 10 quintals of berries per *quindi*, providing five quintals of processed beans, which was selling at 1000 birr/quintal in that year. In addition, they reported that prices had risen by almost 70 percent in the year before the survey.

KHAT MARKETING AND PROCESSING

In stark contrast to coffee, which government restrictions forbid transporting for marketing outside the district, khat is sold in the free market. The government stays out of the market chain, except collecting taxes at tollgates along main highways between the study districts and market towns. Export taxes are levied at the Dire Dawa airport (most khat is flown out of the area to Djibouti, Somalia and London, as road travel is too time-consuming). According to the estimates of a UK Home Office Intelligence Report, in 1998, over seven tons of khat arrived at London Heathrow every week from Ethiopia and Yemen (Griffiths, 1998).

At the time of the survey, there were four processing centers in Deder town alone where khat from the farms in the area was collected, graded and packaged for the journey to Addis Ababa, to other parts of Ethiopia, to the Dire Dawa airport, and to the border towns with Somalia and beyond. The succulent leaf tips and shoots were removed from the main branches, graded, tied into bundles, and inspected for quality. The workers were paid piecemeal for their work, earning approximately 10 birr per day.

The link between the villages and processing centre was managed by middlemen, often local villagers, who acted in a "freelance" capacity, making deals on behalf of the processing centre buyers. The middlemen came to villages to buy khat directly from the field. Transportation was provided by hired donkeys from the village (sometimes even the farmer's own family donkeys). The purchaser organized and paid for the service. In the best villages, at the peak of the season, with several middlemen working on behalf of the different processing centers visiting every day, competition and prices remain relatively high. There was also a secondary market for lower quality khat that was not suitable for export. This khat, named *"chora,"* the poorer shoots of the plant, was enjoyed by the poorer section of society. The petty trading in this product at local markets is solely the work of women.

KHAT VARIETIES AND SEASONAL PRICE VARIATIONS

The price for khat varies widely over seasons across the year. Fresh tips can be harvested throughout the year, but are much more plentiful after rain when the bush undergoes rapid growth. They fetch over five times the price in January, when the season is dry, than in August, when there had been rains in the preceding months. The high seasonal price variation is accentuated because, as with most vegetables or horticultural crops and in contrast to grains, khat's green leafy tip (the most prized part) does not store well. Low supply at other times of the year causes price rises and tidy profits for those with irrigation. Given the limited amount of land under irrigation, and the profits that can be made, irrigation waters are almost always used for khat.

There is also a seasonal price variation with maize and sorghum, though it is much less dramatic. The poor often sell maize and sorghum immediately after harvest when prices are at their lowest in order to finance debt and meet their cash needs, but then buy again later in the year when prices are at their highest. Those farmers who do not have working capital to hold off sales or production until prices have peaked get trapped in cycles of low returns and often increasing disinvestment or indebtedness.

The diversity of khat variety also adds complexity to crop production and marketing strategies. There are several market brands of khat, but the main three in the Deder and Meta are *hamerkot* (or white khat), *diimaa* (or red khat) and *daalota* (which also has a reddish color). These have different properties with respect to the capacity to resist drought, perishability, desirable quality (which includes taste and potency), market preference, and prices. *Hamerkot* khat is the most prized in the local market for its sweet taste and light inebriating effect. However, it is less drought resistant and more perishable than the *diimaa* variety. Its market is hence almost entirely domestic, with a very high demand within the adjacent districts of Meta and Deder. Because it is less perishable, *diimaa* is exported to Somalia, Djibouti, and Europe. *Diimaa* and *daalota* are said to be more profitable than *hamerkot*, because of their readiness for export markets and the large quantity of leaf that is produced.

RELATIVE PROFITABILITY OF KHAT

The equivalent cash returns measured in value yields[6] from khat and coffee (the main cash crops) combined is on average almost five times more per *quindi* than from maize and sorghum[7] (the main food crops). Thus, even though food crops occupy twice the area of cash crops, khat and coffee still bring in 2.5 times the income of food crops (Table 8.3). A similar study in the Habro district (Feyissa & Aune, 2003) found that a khat-maize intercropping system was 2.7 times more profitable than maize monocropping. Not surprisingly a farming strategy of selling cash crops and buying food crops has developed there, resulting in a steady shift from grain and subsistence-oriented production towards growing more market-oriented crops.

In this study, gross rather than net value yields were reported because detailed measures of chemical use, irrigation, seed and labor costs were not calculated separately for each crop. However, on average, chemical and seed costs are approximately 22 birr for an average farm size of 2.5 *quindi*. Thus, net value yields are approximately 2 percent less than gross value yields.

TABLE 8.3 GROSS INCOME PER UNIT AREA [=VALUE YIELD] FOR MAIN CROPS (BIRR PER *QUINDI*)

	Maize	Sorghum	Food Crops	Khat	Coffee	Cash Crops	Total Crop
Low	237	236	235	1312	1068	1217	567
Mid	303	245	282	1279	998	1190	624
High	203	222	206	1228	0	1212	473
Impov	215	209	213	1169	993	1103	515
Poor	255	226	244	1257	1022	1181	556
Better	257	268	254	1363	1135	1331	594
Rich	286	281	263	1489	1014	1337	608
Total	252	232	242	1277	1018	1198	558

In terms of the weight of food grain, instead of value, our survey results suggest an average yield of 2.1 quintals/*quindi* (about 1.7 tons per hectare). Our qualitative data generated from discussions with farmers supports these general finding. In three *kebeles*, farmers reported that yield (assuming no drought) ranged from as low as 0.5 to 1 quintal/*quindi* for the poorest land, using traditional seed with no purchased inputs and up to 5 to 7 quintals/*quindi* for hybrid seeds and inorganic fertilizers on good flat soil owned by rich farmers. This amounts to a range of 0.4 to 5.6 tons/hectare.

An examination of the crops giving the highest returns shows that Irish potato earns the most per *quindi* (1610 birr), followed by khat (1280 birr) and closely by coffee (1020 birr).

Maize, groundnuts and sorghum all fetch between 230 and 250 birr per *quindi*, while wheat fetches only 70 birr per *quindi*.

A similar range in returns was reported in focus group discussions with farmers who grew khat. Returns varied from about 500- 1000 birr/*quindi* for those who didn't irrigate or compost, and up to 5000- 8000 birr/*quindi* for those who irrigated right through the dry season and gave high attention to weeding and mulching. This variation is due in part to the very high 'off-season' prices that can be obtained for khat, but it is due in part to much higher management and irrigation inputs.

THE CONTRIBUTION OF KHAT INCOME TO RURAL LIVELIHOODS

This section looks at incomes from khat sales, association between khat production and intensification of food crop production, and livelihood graduation trajectories in the non-farm sectors. These analyses give a fairly clear picture of the contributions of the khat enterprise to livelihoods of the producing households.

On average the khat yield of rich households is over a quarter more than the impoverished households. The yield amount rises to a third more for maize and sorghum. We find that the richest households use 9 times as much inorganic fertilizer and 24 times the value of improved seed for food crop production as the poorest households. Rich households are also 3.5 times more likely to have high quality (as opposed to moderate or no) irrigation access and they use seven times more hired labor, primarily for khat activities.

The findings of the current survey are consistent with the results of earlier empirical investigation conducted in Haramaya, Kuni and Sabale districts. According to Tefera et al. (2005), the revenue generated from khat sales accounted for 28.7 percent of the average annual income of farm households and about 85 percent of total income from cash crop sales in the three districts. The authors also reported that there were synergies between cash cropping (khat) and the intensification of staple crops, maize, and sorghum production. The income from sales of khat was used to finance the adoption of inorganic fertilizers and improved seed-based intensification of staple crops production and also to finance grain deficit. In addition, Tefera et al. (2003) found positive and strong correlation between the extent of households' participation in commercial khat production, particularly irrigated khat, and food security and pre-school children's long-term nutritional status in Haramaya, Kuni and Sabale.

TABLE 8.4 LIVELIHOOD POVERTY OPTION MAP

	Cash crops	Food crops	Livestock	Non farm
Impoverished	Unirrigated khat/coffee	Unfertilized maize, sweet potato	Chicken & eggs selling	Agricultural labor work, particularly migratory, Firewood collection
Poor	Unirrigated khat/coffee	Fertilized maize and some irrigation	Goats rearing	Goat trading
Better-off	Partly irrigated khat	Fertilizer and irrigation	Cattle fattening and donkey renting, milk sales	Khat & grain trading
Rich	Irrigated khat and Irish potatoes	Fertilized and irrigation, Hire labor and use improved seeds too	Oxen hire and dairy	Shop, Pump renting remittances

Further, the current analysis shows the livelihood portfolios of the poorest farmers remain dominated by low return agriculture because they have not been able to diversify into more profitable livestock to start accumulating via the livestock ladder. On the other hand, rich households often specialize in crop cultivation because, if good irrigation and market access are available, it gives them the highest returns on their investments compared to the non-farm sector (Table 8.4). In contrast, the impoverished men augment their income through employment in agricultural labor, particularly outside the village, while their wives are supported by the public Employment Generation Scheme (EGS) and income from firewood collection and selling.

The next group up, the poor households, begins to get more involved in goat and vegetable trade and non-agricultural labor. The better-off households predominantly move into khat and grain trade, which requires reasonably good access to khat, grain and working capital. Meanwhile the rich households move into such businesses as convenient stores, retailing consumables, pump renting, and earning good remittance from their educated sons, sometimes sent abroad as a deliberate livelihood diversification strategy.

THE CONTRIBUTION OF KHAT INCOME TO RURAL ECONOMY

The discussion has thus far focused on those with access to irrigation. The survey result indicates that only 30 percent of the farm households have some direct access to average quality irrigation, so it is only a limited number that can benefit directly from any crop diversification. Any intervention to increase irrigation takes many years. What happens to those without it in the meantime? What is the impact of the irrigation sub-sector on the wider economy and to what degree do funds circulate back into other sectors that advantage the poor?

It is clear from the current analysis that the irrigated crop economy—mainly based on khat—does have profound effects throughout the wider economy. The limited scope of our study did not allow for a more comprehensive analysis of local money flows within this study, but a simplified analysis of expenditure flows from those households who are directly involved in producing export cash crops suggests that cash cropping has relatively high local multiplier effects. In this regard, the discussions conducted with khat farmers on their expenditure patterns enabled us to analyze these money flows.

Khat money is recycled locally by buying-in goods and services from others in the villages. In addition, there are indirect linkages to information flows, knowledge sharing, and market development, to name a few. Ultimately wage rates, and the availability of labor work, also depended on how well the khat economy is doing at a specific time.

A rich farmer may earn about 1500 birr net from khat in a year. In gross we can estimate this at about 2000 birr. The trader pays for about 20 donkey trips, at 7.5 birr/day (=150 birr) to transport khat to the market. The household will then spend about 200 birr employing one person for about 30 days a year and may spend the same amount on pump renting for irrigation at the average rate of 200 birr. The household buys food stuffs, such as eggs, chickens and milk in the village to the value of another 150 birr and goat meat to the value of 150 birr. There might also be investment in house improvements using local labor and materials, such as red soil (100 birr), plus expenditure on collected firewood and fodder (75 birr). Overall 51 percent of a farmer's gross income of 2000 birr from khat (1025 birr) gets recycled immediately into the local economy, of which 725 birr goes directly to women, and 600 birr goes to the poorest farmers. These groups in turn recycle their money locally buying goods and services from others in the villages. In addition there will be indirect linkages through information flow, knowledge sharing, and market development and so on.

CONCLUSION

The empirical evidence presented here shows financially well off households in the study area are those who own irrigated khat fields. The income from khat sales is used by these households to finance the intensification of food crop production, through the use of improved seeds and inorganic fertilizers. Khat producers are able to make up for grain deficits through purchases, and thus khat producing households have a better food security status. Moreover, analysis of expenditure flows from khat growing households shows khat money is recycled locally on purchases of goods and services from others in the villages. In particular, wage rates and the availability of jobs depend on how well the khat economy is doing.

Coffee remains an important crop. It is showing signs of price recovery, but it is less important than khat, which maintains price advantage. The equivalent cash returns from khat and coffee are on average five times more per *quindi* than from maize and sorghum. Consequently, almost all households in Deder and Meta have increased their khat cultivated area at the expense of food crops. The increases have applied to households of all wealth groups.

In a nutshell the current analysis confirms that improved access to market, increased demand, and price incentives for producers have been the primary drivers of the khat-based agricultural commercialization in eastern Ethiopia. Low costs of production, low production risk, except for frost, and a liberalized market have made khat production the most profitable enterprise. The increasing shift to khat production in Deder and Meta has been partly a response to context-specific opportunities and challenges. The main factor is of course the fact that the districts are a part of the Harerge highlands which are characterized by drought-proneness, high population pressure, steep slope, and hence low land productivity and income. The removal of vegetative cover and cultivation on steep slopes are the main causes of soil erosion and land degradation in the area.

Under such conditions, meeting the increasing subsistence needs makes intensification, a strategy that must of necessity yield high return per unit of scarce land/water, an inescapable imperative, but at the same time, requires low external input and entails low risk. This is a challenge that currently only khat has been able to meet. Planting khat hedgerow is the most preferred and economically attractive conservation method providing year round vegetative cover. Unlike coffee, khat needs no agro-chemicals and the cash crop is rarely affected by diseases; it can tolerate several months of moisture stress; and the practice of intercropping khat with food crop on steep slopes ensures seasonal income flow and an even distribution of labor requirements. Further, the crop can grow under a variety of climatic conditions. As a perennial crop, khat is an important means for accumulation, intergenerational asset transfer and old age insurance, in a situation where land ownership belongs to the state. Moreover, the effects of the shift to khat on income, food security and poverty have been found to be positive, with locally recycled khat income generating positive multiplier effects in the rural economy.

Despite all the benefits, khat is not a panacea for the ills of smallholder agriculture. The crop is susceptible to frost damage and has a variable market price. Unlike annual crops, it does not allow for flexibility in land use. At least a five year time horizon is needed when planning cropping strategies, which reduces farmers' ability to adapt and respond to changes in market situations or environmental change through flexible resource use. Moreover, unchecked expansion of khat production against its alleged adverse health effects and its status as a controlled substance in some countries cast a shadow on the sustainability of khat-based livelihood strategy and national export earnings.

From a policy standpoint, the dilemma lies in officially recognizing khat's roles and providing extension support for a crop considered illegal in some countries. On the other side, creating alternative viable livelihood options within the context of the densely populated, environmentally degraded, and drought-prone eastern Ethiopia is a formidable challenge.

Of necessity, due consideration must be given to potentially devastating damage that the policy of hastily 'criminalizing' khat producers and traders could have on the livelihoods of the poor and on the struggling Ethiopian export sector. Focusing efforts on restraining the growing demand for khat through effective communication, information and education programs is probably a more feasible strategy in the medium term. In the long run, a concerted and collaborative effort is indispensable to encourage and support khat producers and others, particularly through meaningful development initiatives, to enable them to gradually reduce their dependence on khat through diversification to other farm and non-farm activities.

Indeed a rural livelihood analysis which was conducted in the two districts[8] has highlighted that there exists a potential for the development of horticultural crops in the long term, perhaps using out grower schemes, for those villages closer to road, and for those households with a better access to irrigation. The better-off farmers in the villages do have the potential to develop skills in new crops if further improvement is made to the market infrastructure. Thus, interventions might be important to support further irrigation development, yield growth, crop diversification, access to credit and marketing. For a low potential area, an important strategic option might be identifying areas where irrigation could be developed, but supporting soil and water conservation, developing alternative income streams (such as bee keeping, poultry and forestry), providing support to seasonal and permanent migration and direct transfers to the poorest in times of hardship.

NOTES

1. The fieldwork for this chapter was conducted as part of rural livelihoods analysis in Deder and Meta districts – a consultancy assignment commissioned by Oxfam GB/ Ethiopia. We thank all Oxfam staff. Our special acknowledgement goes to Petra Kjell of the New Economic Foundation (NEF) and Mulu Tesfaye for their crucial inputs to enriching the livelihood analysis. The first author is grateful to Professor Belay Kassa, President of Haramaya University, for his encouragement and for creating an environment for the professional enhancement of young scientists,

which in turn enabled them to make a better contribution to the national development agenda.
2. Lower administrative units covering up to 13 villages.
3. Sub *kebeles* of two to three villages.
4. The results presented in the tables are those generated by the authors' field research and analysis, unless otherwise indicated.
5. Birr is local currency, exchange rate was about 8.7 birr for one USD at the time of the survey.
6. This is defined as income per *quindi* of land.
7. This value reflects food crops consumed with their potential values in the market at an average farm gate sale price of 130 per quintal (100kg) of maize and sorghum.
8. See Start et al. (2005) for further details on rural livelihoods in Deder and Meta.

PART III

POLITICS OF POLICYMAKING

Chapter 9

Beyond the Politics of Prohibition

Ezekiel Gebissa

ഗ‍ര

Among the many questions pertaining to khat and its use, the chapters in this volume have focused on the impact the expansion of khat cultivation and the high revenues from its sales are having on Ethiopian economy and society. There are concerns about increasingly more farmland being allocated to the production of cash crops in a country frequently beset by food shortages, the sustainability of an agricultural system dominated by water-intensive perennial crops, and the potential for the development of a narco-society if the current rate of expansion of the habit of khat chewing is allowed to continue unchecked. While these challenges are real, the chapters have argued collectively that prohibition is not a practical solution. Khat represents a national dilemma. Its multifaceted functions are too complex to be reduced to boilerplate good-versus-bad articulations of the impact of khat. By bringing into focus the arguments presented in the preceding chapters, this chapter juxtaposes the benefits and challenges of khat and outlines a framework for a debate leading to the formulation of an effective policy regulating khat use, production, and trade.

Khat has direct micro and macroeconomic effects as a source of substantial tax revenues for regional governments and foreign exchange earnings for the national government. The industry creates employment for the people associated with its marketing and generates income for entities like the Ethiopian Airlines which transports the export freight. It is now established that growing khat has environmental impact on soil conservation while allowing farming families to maintain food self-sufficiency, satisfy their commodity needs, and meet their various cash obligations. Khat-generated financial prosperity and increased transportation facilities have engendered a number of improvements in the quality of life of the farmers who grow the shrub. The khat industry has brought a number of commodities, for example, mattresses, blankets, clothing, household utensils, and building materials to farming communities and made them available at affordable prices. These commodities have made life easier, relatively comfortable and healthier for those who derive their livelihood from khat related economic activities.

Notwithstanding the contention of many experts, converting cereal fields to khat orchards does not necessarily portend food insecurity for the farmer. In contrast to other cultivated perennial cash crops, such as coffee, khat in fact enables farmers to grow food crops through intercropping, particularly during the "establishment stage." Food or other crops typically inter-cropped with khat include beans, sweet potato, vegetables, sorghum and corn. Because of its agronomic and price advantages over other crops, researchers suggest that the economic incentives of khat cultivation and trade would remain very high (Storck et al., 1991; Stage & Rekve, 1998; Poschen, 1986). As a cash crop, khat is very attractive to farmers as it generates very high income with minimum production input. Furthermore, demand for khat has continued to increase and farm-gate prices have accordingly continued to rise. Under these conditions it is unrealistic not to expect expanded cultivation of khat in the future.

The fact that demand for khat is on a steady increase suggests a relatively elastic market for the commodity domestically. Yet it is impossible to imagine Ethiopia forging ahead with a non-food luxury product for which there is no reliable international demand. Khat may never become a legal export commodity to countries that can afford to pay premium price and sustain its market. It is already a controlled substance in the United States, Canada, and several countries in Europe. Wherever it is legal in a Western country, the market is among the immigrant communities from the Horn of Africa. A realistic option for an expanding market is the development of a khat-based pharmaceutical product or an ingredient in drinks. However, attempts at converting khat into a popular product have never caught the Western public's imagination. Khat would likely remain a regional commodity with little prospect of an expanding international market.

Despite the fact that its impact is regional, khat faces strong international opposition. In reality, that the level of cultivation and use is already extensive makes prohibition difficult. It has also been shown that the use of legal measures to limit the alleged damages to consumers or the problems associated with khat as an agricultural commodity, such as the purported social problems of chronic use and the potential displacement of food crops, is ineffective. In the 1920s and 1950s, colonial officials in Djibouti, the Aden Colony, British East Africa, and British Somaliland rushed to ban the cultivation and use of khat citing the socioeconomic adverse effects. The measures were met by stiff resistance, forcing the governments to rescind them in all of these countries. Here it is worth noting that the argument about the adverse effects of khat originated with the British and French colonial officials who had inadequate knowledge about khat and harbored legendary bias toward local customs and traditions.

It must also be emphasized that the situation in Harerge, the region where the khat cultivation that now characterizes agriculture in several parts of Ethiopia began, has started to exhibit signs that signify the precariousness of a khat-based agriculture. The economic conditions which obtained in Harerge during the

1990s, a period of relative prosperity for most khat farmers may turn out to be only a fleeting moment of plenty unless innovative policies are put in place to ensure sustainability. There are signs that the boom and its consequent benefits are tapering off in some places, as shown in chapter 7. Given the shrinkage of land holdings, it is unlikely that the standard of living which the khat boom has engendered would be sustained for an indefinite period. Thus far, the income that flows from khat has relieved the state of the obligation of agricultural extension costs and assured an orderly migration of a few farmers to non-farm occupations and to life in urban centers. Unless some alternative valuable export commodity is found to replace khat or farmers are shown alternatives to farming as an occupation, the improvements that farmers have experienced in their lives are very likely to be lost and farmers forced back into subsistence life.

Just as the impact of banning could be disastrous, the damage that khat chewing may cause in a country where health and social welfare programs are non-existent or poorly developed is not something to be taken lightly. Officials are obligated to encourage the population to make considered choices to minimize the negative consequences of khat misuse. The dilemma in this regard is the fact that khat chewing on social and religious grounds is deeply rooted in the socio-cultural fabric of the communities where khat is chewed in the traditional setting. The studies on the medical, psychological, and pharmacological impact of khat on the user have not been conclusive enough to form a consensus as to the effect of khat. Armstrong (2008), Odenwald (2007) and Al-Hebshi & Skaug (2005), have shown the inconsistencies, contradictions, and tentativeness in the literature. The prohibitionists do not need to overdramatize the negative effects of chewing. Educating the public on khat is thus vital to attaining positive and voluntary changes gradually by the society since any change imposed by law enforcement in areas where there is a large consumer population will be resisted equally by both the producers and the consumers. In the end, we must realize that policy decisions regarding restriction would be made largely on economic considerations. In other words, there is little chance of government intervention to proscribe khat use in the near future.

PRODUCTIVE DEBATE AND EFFECTIVE POLICY

In the regions where khat chewing has a long history, its use is not only well integrated into the social life of the communities but it also fosters this integration in the present. It pervades every aspect of life so much that it is impossible to explore, in the words of Bornislow Malinoviski, "the hold life has" (Allen, 2002) without understanding the place the use of khat has in the life of the people of Ethiopia. In Harerge in particular, the indigenous uses of khat are so fundamental that there is hardly an important event in the life of an individual in which khat does not play a crucial role. Here it is unlikely to be abused because of the limits that cultural forces surrounding the chew events impose on the amount ingested. In this sense, the uses that remain close to the formalities of the indigenous context must be viewed separately from the contemporary chew practices where the opportunity for misuse is greater. Only individuals who chew outside the traditional context are likely to misuse khat. And for those chewers outside the traditional context, only moral suasion can be expected to be a more effective tool to keep them away from misuse.

Thus far, the controversy over khat use has ignored such important nuances and distinctions as the difference between khat chewing in the traditional context and the "modern" setting. Any effort to articulate the socioeconomic and cultural significance of khat is often characterized as a defense of an indolent behavior. This tendency trivializes the debate and undermines the possibility of formulating a sensible, evidence-based khat policy. The prohibitionists claim moral superiority for trying to save the chewers from the alleged ravages of their own culture. This is an untenable position because it is practically impossible to eradicate the practice of chewing without having to require the chewers to reject their culture in the process of saving themselves ostensibly from a harmful habit. Because opponents rely on realities that are foreign to the chewers and producers, such as the effect of chewing on sperm motility index, their effort to restrict, let alone eradicate, consumption have not made a slight dent so far. In fact, the evidence shows that khat chewing has become a pan-Ethiopian phenomenon. If the goal of prohibition is to extirpate the harmful aspects of chewing, then the built-in cultural mechanisms of moderation should be recognized and strengthened rather than attacked and destroyed. The experience of other countries shows, despite the documented importance that cultural factors play in controlling substance use and promoting harm reduction, legislation and criminal justice approaches focus solely on the alleged harmful and deadly effects and seek completely to eradicate the use of the substance misleadingly labeled as a drug. There is little evidence in history or in contemporary endeavors that the 'combat' approach has ever delivered adequate results.

In debates about drug control, arguments that articulate the view of the farmers and indigenous chewers are often lumped together with the "dangerous" category that includes the so-called promoters or "drug pushers" and dismissed out-of-hand (Zinberg 1986, p. xii). In Ethiopia, this kind of guilt by association exists. It may serve as a powerful tool to scare the public into supporting the prohibitionists' position but it does not advance the cause of harm reduction. There are, of course, those who promote the indiscriminate use of the leaf in order to exploit the producers and augment their own economic interests. What is missing in the discourse on khat in Ethiopia is a genuine exchange of ideas that can inform the process to formulate a workable regulatory policy that can address the legitimate concerns of both the proponents and opponents of restrictions or eradication. Recognizing the positions of the other as principled can go a long way in resolving the controversy and in developing a national policy that can effectively prevent misuse.

Despite its obvious benefits, khat production could adversely affect rural life in three main areas. First, unrestricted expansion of khat production precludes sustainable development if the current actions of the producers compromise "the ability of future generations to meet their own needs" (WCED, 1987, p. 43). One cannot envision a scenario whereby the problems associated with khat production, primarily the reduction of land resource and the rapid overpumping of stored water for irrigation, could have a positive effect on the environment in the long run. A related issue is income sustainability. Khat has brought great benefits to the villagers engaged in growing it, but the prosperity can only be viewed as ephemeral. The income farmers derive is not being reinvested on the essentials of economic development but spent on consumer goods. The farmers have only a few ideas of viable investment opportunities.

Second, the present prosperity is lopsided. While on the one hand many households are prospering because of their income from khat, on the other hand, low-and mid-income chewers are certainly contributing to their own impoverishment. Regular chewers spend a third to one half of their income on khat. The critics of khat argue that the prosperity enjoyed by one sector is the cause of impoverishment of another. But they are not justified recommending the banning of khat production. While the remedy to the social problems of compulsive spending on khat is individual responsibility, the lopsidedness of the prosperity is detrimental to the sustainable development that Ethiopia as a country aspires to achieve.

Third, while agricultural intensification is perhaps the way to proceed, the increasing trend toward cash cropping nationwide could lead to more and more areas of land being locked up in tree crops and the simultaneous stress brought to bear on finite water resources. If this trend continues, national food production would diminish, since it is always the best, most productive lands that are converted to khat orchards. The country would be forced to spend ever larger sums of hard currency on food imports just to maintain the status quo, exacerbating the already fragile food condition in the country. Khat is unquestionably beneficial to the households that grow it, but a khat-based economic strategy cannot be the basis for a sound policy for a sustainable development.

The national debate on khat thus needs to move away from sterile exchanges about the deleterious physical and social effects of khat on one side and the cultural/economic benefits on the other to the more important debate about the future. This means that a shift in the public discourse needs to occur, focusing on the three main elements of sustainable development: ***people*** or social aspects, ***profits*** or economic benefits, and the ***planet*** or environmental concerns (Rafn, 2002). Regarding the first element, some policy-makers have advocated the inclusion of the views of consumers, farmers, traders, and other auxiliary service providers whose livelihood depends on khat in the national debate. From the vantage point of the critics, "none of these sectors is expected to approve the banning of [khat] in the country" (Woldemichael, 2004). In other words, the advocates of a complete khat ban do not want anyone connected with the khat industry to be a part of this national discourse. Instead, they prefer a draconian measure implemented by the coercive arm of a state presumed to be benevolent or at least one which is assumed to know what is in the best interest of the people.

In fact, the government should not be supposed to know the interest of khat farmers more than the farmers themselves. That is why some observers urge caution, suggesting that

> more research needs to be done before rash decisions are made and livelihoods ruined. They say that the best knowledge comes from Ethiopia's khat farmers and advise that the government should works [sic] more closely with khat farmers to define a clear policy whether khat really is productive for Ethiopia's long-term development" (Bhalla, 2002, Youth Problem Section, para. 10).

Put another way, those who are directly involved with khat should be included rather than excluded from the debate. Given the economic importance of khat, only a public discourse that accepts the legitimacy of the position of the defenders of indigenous uses of khat can lead to a resolution of the khat controversy.

The remaining two aspects of sustainable development are equally important. After considering both sides of the khat issue, Mulatu and Kassa (2001) have come to a considered conclusion that "despite its reported negative effects on health and other social aspects, khat was playing important economic and possibly ecological roles for farmers in the Harer Highlands of Ethiopia" (pp. 106–107). In addition, Klingele (1998) considers khat the only agroforestry plant that could stabilize erosion control structures while yielding high market value that allows for a stable and substantial cash return. In an area of high population density and rapid growth rate, farmers cannot afford to lose scarce arable land to erosion control structures and khat fills "the gap in a very profitable way no other plant could do" (Klingele, 1998, p. 17). In other words, at least at present, khat plays a unique role concerning the profit and planet imperatives of sustainable development. The real issue is not "to ban or not to ban," but how best to use the present khat-generated prosperity in a way that will also benefit the next generation.

Chapter 10

Afterword

Christopher Clapham

ಸಿರ

Khat cultivation provides a striking example of the ability of Ethiopian smallholders to respond dynamically and imaginatively to the opportunities presented by regional and international markets. In a country notorious for famine, in which increasing population and environmental degradation are often regarded as imposing a downward Malthusian spiral of relentless pressure of hungry mouths on declining resources, they have found a way not merely to arrest but – for the present, at least – to reverse the decline in living standards, and to provide for themselves and their families opportunities of which they could once only have dreamed. Stony, steep and apparently valueless hillsides are neatly terraced, and planted with bushes that not only prevent erosion, but provide an income unavailable to those who toil at the production of sorghum or millet. As Tesfaye and Start argue elsewhere is this volume, khat is the only economically viable crop that also stabilizes hillsides. Lorries, pick-up trucks and aircraft speed the leaves to waiting markets, not just in the Horn of Africa and nearby Arabia, but increasingly across the world. In the way that classical economists from Adam Smith onwards have recognized, specialization in an area of production in which they enjoy a comparative advantage enables khat farmers to escape from the straitjacket imposed by the demands of self-sufficiency, and tap into the global economy to make good their need for products that they could only create at far greater comparative cost than more favored producers elsewhere. Khat cultivation likewise requires a significant level of capital investment on the part of smallholders, both in the construction and maintenance of terraces where these are needed, and in the waiting period of about five years before a newly planted bush can pay for itself. It is hard to imagine a clearer refutation of that now mercifully discredited stereotype that once saw 'peasants' in general – and Ethiopian peasants in particular – as locked into age-old systems of production that excluded the possibility of change, and from which they could be rescued only by benevolent governments and aid donors equipped with the 'modern' technologies and ideas required to bring them the benefits of 'development'.

Paradoxically, indeed, it is the government and donors who warn against the dangers of specialization, and seek instead to promote the merits of self-sufficiency. The idea that smallholders especially might engage in farming strategies that required them to purchase food, rather than growing it themselves, is evidently deeply worrying – perhaps understandably so, in a country that has suffered as badly as Ethiopia from shortages of basic foodstuffs. The agricultural policies of successive Ethiopian governments, from the Chilalo Agricultural Development Unit (CADU) under the imperial regime in the 1960s, through the state farms and producer cooperatives of the Derg, to the Agricultural Development Led Industrialization (ADLI) policy of the present EPRDF regime, have been predicated on the need to grow more grain. By 'taking the place of food', both at the level of production by using land and labor that might otherwise have been devoted to growing cereals, and in some degree at the level of consumption, given that khat is chewed amongst other benefits for its appetite-suppressing qualities, the leaf appears to strike at a very basic idea of what farming in Ethiopia 'ought' to be about.

Afterword

There are distinctive features of the crop that enable smallholders to capture a larger share of the surplus generated by khat cultivation, in contrast notably to alternatives such as grain and coffee. Most important is the inability – as yet, at least – to stabilize and refine the volatile active element in the leaf, cathinone, which as a result can be accessed only by chewing, within a maximum of two days after it has been picked. This peculiarity, which militates strongly against the commoditization and commercialization of khat, in practice operates to the benefit of its producers. Notably, it virtually excludes the imposition of heavy-handed government control, of the kind that has suffocated production of most other commercial crops. Even the Derg regime, which imposed punitive farm-gate prices on coffee producers, and still more regressive quotas on grain farmers, was quite unable to develop equivalent surplus-extraction mechanisms for khat, which as a result flourished at the expense of tightly-regulated alternative crops. Though the EPRDF government's ADLI policies are notably less repressive than its predecessors', they still impose a political cost, through effective state control (often exercised through 'partystatal' corporations owned by the regime) of the supply of vital inputs such as improved seed and fertilizer, which have likewise, as Degol Hailu argues elsewhere in this book, given a comparative advantage to khat production. Khat farmers, like the producers of other crops, are certainly subject to potential exploitation by purchasing agents, but the existence of at least a relatively free market, coupled with a high level of local consumption that gives them an alternative outlet for their leaves, provides them with a measure of protection. Price volatility for khat also appears to be relatively low, compared with that for alternative products. I have not seen any figures comparing returns to khat producers as a percentage of the final selling price with those for other crops, but would expect these to be markedly higher.

The other distinctive feature of khat is its ambivalent position on the borderline between legal and illegal psychotropic substances. The global market for, and trade in, psychoactive products of one sort or another is massive, and includes very high levels of trade in substances which despite their psychoactive properties are perfectly legal, such as alcohol (except in a number of Moslem states), coffee, and tobacco. This trade is characteristically dominated by large multinational corporations (British-American Tobacco, Distillers Corporation, Nescafe), and subject to high levels of commodification, with associated product differentiation, advertising campaigns, and other elements of the global marketplace. At the other extreme, highly refined narcotics such as opium and stimulants such as cocaine are illegal virtually throughout the world, and though they too generate massive markets, these operate undercover through criminal networks and other organizations (such as guerrilla movements in parts of South America and southeast Asia) which flourish precisely through their ability to evade formal state and international control. The khat market, by contrast, operates outside either of these systems. The fact that the crop is illegal in much of the developed world must play a significant role (along with the volatility of the alkoloid, as noted above) in deterring global corporations from becoming involved in its production, distribution and marketing. Companies that market tobacco, for instance (albeit under increasing restriction), could scarcely be expected to take over the marketing of khat, which would place them under the onus of trading in drugs on the borderline of legality. It remains an open question whether the global legalization of khat would benefit producers, by leading to the development of more efficient distribution and marketing systems and greatly increasing the scale of the potential market, albeit under the control of major corporations.

There can scarcely be any doubt, however – and this is something that the contributions to the present volume make overwhelmingly clear – that the effects of any global prohibition on the production and marketing of khat would be disastrous, not only for the producers but much more widely. For a start, as Ezekiel Gebissa's *Leaf of Allah* has demonstrated, numerous attempts by different government authorities in the Horn of

Afterword

Africa and Arabia to prohibit or control the production and distribution of khat have proved entirely ineffectual, merely demonstrating their own impotence in the face of the interests and ingenuity of producers, distributors and consumers alike. The only measure likely to have any discernible effect would be the wholesale eradication of the trees themselves, and this in turn would result in the instant impoverishment of very large numbers of people, and the intense political alienation of both producers and consumers, in an extremely volatile region of the world already deeply affected by militant Islam and the reactive (and often over-reactive) 'global war on terror.' Eradication would not strengthen the state, but weaken and possibly destroy it. It would also have a sharply negative effect on government revenues. Though regional Islamists have generally opposed khat consumption, as antithetical to their vision of a purified Islam, it is hard to believe that they would not find some way of tapping into the alienation that any serious attempt (by anyone else) to control khat consumption would create.

This is not to deny that excessive khat consumption does indeed carry significant social costs. Other contributions to this volume have described the large amounts of time spent indolently in chewing sessions, and the amount of income of often very poor people dissipated on buying the essential leaves. Any visitor to areas of large-scale khat consumption can testify to its longer-term debilitating effects in reducing some men (by far the main consumers) to premature senility. Yet virtually every society in the world has access to, and places heavy dependence on, mild psychotropic substance of one sort or another, the overconsumption of which – alcohol, in particular – may be every bit as damaging as that of khat. The worldwide evidence is that the deleterious impact of mild psychotropic is far more effectively managed by legalization accompanied by social sanctions on abuse, rather than by outright prohibition, and that attempts by governmental authorities to control abuse can work only when backed up by wider social support.

In this respect, the inability to stabilize and refine cathinone is enormously beneficial. One need only compare khat with another leaf, coca, that has for generations been consumed by the peoples of its growing areas in South America under conditions broadly comparable to the traditional uses of khat, but that has been transformed by its use as the basis for a highly refined product, cocaine, which has not only destroyed the lives of many of its consumers, but has turned its areas of production into battlefields between criminal gangs, guerrilla insurgencies, and the often corrupt repressive forces of the state. Indigenous peasant producers, ostensibly the beneficiaries of the crop, must in practice be counted amongst its victims. Much the same could be said for the relationship between poppies and opium across a swathe of Asia. Were cathinone to be 'captured', and refined into a hard drug with a global market equivalent to opium or cocaine, the impact on its zones of production would be terrifying.

At the same time, though the arguments against any general prohibition on khat appear to me to be unassailable, it would be a mistake to go to the opposite extreme, and extol the crop as a potential basis for any broader process of sustainable development. Khat has certainly helped to mitigate the impact of intensifying population pressure on resources in its main areas of production, and has generally, as argued elsewhere in this book, improved rather than undermining the food security of its producers. It has likewise had beneficial multiplier effects in providing start-up capital and economic opportunities for small business such as transport, enabling people to gain access to imported products that have materially improved their lives, and meeting the costs of education. The khat boom must nonetheless be subject to many of the same dangers that have affected other primary product booms in other parts of Africa – cocoa in West Africa, for example – and may well prove short-lived. Tree crops such as khat certainly have some advantages over annual crops such as grain, notably in their capacity to withstand drought years, and their low labor demands once they are established; but they also carry risks, notably that the long lead time between planting and production may lead to over-investment in planting when times are good, and consequent over-supply and price collapse by the time trees are ready for cropping. In short, though khat has a very welcome role in maintaining and even improving the livelihoods of smallholders in one of the poorest areas of the world, it offers no escape from the much larger problems of African agriculture.

Bibliography

೩⃝ಌ

Abebe, D., Debella, A., Dejene, A., Degefa, A., Abebe, A. & Urga, K. (2005). Khat chewing habit as a possible risk behavior for HIV infection: A case-control study. *Ethiopian Journal of Health Development* 19, 3, 174-181.

Adinew, B. (1991). *The Analysis of Land Size Variation and its Effects: The case of small holder farmers in the Harerghe highlands.* M. Sc. diss., Department of Agricultural Economics, Alemya University of Agriculture.

Adinew, B. (2005). The Economic Impact of T'chat Production and Marketing: a dilemma of short term benefits and long run costs. *The National Conference on Khat Habit and Other Psychotropic Drugs and the Spread of HIV/AIDS on Health and Socieoeconomic Wellbeing.* 23-24 May 2005. Addis Ababa: Ethiopian Health and Nutrition Research Institute.

Adugna, F., Jira, C. & Molla, T. (1994). Khat chewing among agro high school students in Agaro, South Western Ethiopia. *Ethiopian Medical Journal, 32*, 161-166.

Advisory Council on the Misuse of Drugs (ACMD). (2005). *Khat (Qat): Assessment of risk to the individual and communities in the UK.* London: Home Office.

Ahmed, H. (1988). Introducing an Arabic Hagiography from Wallo. In Taddese Beyene (ed.), *Proceedings of the Eighth International*

Conference of Ethiopian Studies (pp. 185-97). Addis Ababa: Institute of Ethiopian Studies.

Ahmed, H. (1997). "A Brief Note on the Yemenis of Ethiopia" In K. Fukui, E. Kurimoto & M. Shigeta (eds.), *Ethiopia in Broader Perspective: Papers of the 13th International Conference of Ethiopian Studies*, vol. I (pp. 332-348). Kyoto: Shokado Book Sellers.

Ahmed, H. (2005). "*Waqf*-land in Nineteenth-Century Wallo (Ethiopia)" In Donald Crummey (ed.), *Land, Literacy and the State in Sudanic Africa* (pp. 241-50). Trenton: The Red Sea Press.

Ahrens, J. D. (1998). *Food Shortages Force Oromo of East Hararghe into Migration*. United Nations - Emergencies Unit for Ethiopia, Addis Ababa.

Alem, A. & Shibre, T. (1997). Khat induced psychosis and its medico-legal implication: a case report. *Ethiopian Medical Journal*, 35, 2, 139-9.

Alem, A., Kebede, D. & Kullgren, G. (1999). The prevalence and socio-demographic correlates of khat chewing in Butajira, Ethiopia. *Acta Psychiatrica Scandinavica. Supplementum*, 100, 397, 84-91.

Al-Hebshi, N. & Skaug, N. (2005). Khat (Catha edulis)--An Updated Review. *Addiction Biology*, 10, 4, 299-307.

Ali, H. (1985). *A short biography of Bejirond Takla Hawariat*. Senior essay. Department of History, Addis Ababa University.

Allen, C. J. (2002). *The Hold Life Has: Coca and Cultural Identity in an Andean Community*. Washington: Smithsonian Institution Press.

Anderson, D. & Carrier, N. (2006). Contested Narrative of Khat Consumption. In John Brewwer and Frank Trentmann, *Conusming Cultures, Global Perspectives, Historical Trajectories, Transnational Exchanges* (pp. 145-166). Oxford & New York, Berg.

Anderson, D., S. Beckerleg, D. Hailu, & A. Klein. (2007). *The Khat Controversy: Stimulating the Debate on Drugs (Cultures of Consumption)*. Oxford & New York: Berg.

Anonymous. (1945). The Need for the Control of Khat. *East African Medical Journal*, 22, 1, 9-10.

Bibliography

Anonymous. (1945b). Editorial—Poisoning by Khat or Miraa, Catha Edulis. East African Medical. *East African Medical Journal, 22,* 1, 1-2.

Appadurai, A. (1996). *Modernity at Large: Cultural Dimensions of Globalization.* Minneapolis: University of Minnesota Press.

Armstrong, E. G. (2008). Crime, Chemicals, and Culture: On the Complexity of Khat. *Journal of Drug Issues, 38,* 1, 631-648.

Arrafaine, M. (2004). *Micro-macro economic analysis of "Khat" production in Ethiopia.* M.A. thesis. Department of Economics, Addis Ababa University.

AUA, Alemaya University of Agricultre. (1986). *Retrospect and prospects of agricultural research and extension at AUA.* Diredawa: Alemaya University of Agriculture.

AUA, Alemaya University of Agriculture (1985). *Project Appraisal Document for Alemaya Woreda.* Dire Dawa: Farming Systems Research Unit.

Awasa Finance-Bureau (1996). *The Khat Improvement Law (in Amharic).* Finance Bureau, SNNPR, Awasa, Ethiopia.

Bally, P. R. (1945). Catha Edulis. *East African Medical Journal, 22,* 1, 2-3.

Beckerleg, S. (2008). Khat Special Edition Introduction. *Substance Use and Misuse, 42,* 6, 749-61.

Beke, C. (1843). On the countries south of Abyssinia. *Journal of the Royal Geographical Society, 13,* 254-269.

Belew, M., Kebede, D., Kassaye, M. & Enquoselassie, F. (2000). The magnitude of khat use and its association with health, nutrition and socio-economic status. *Ethiopian Medical Journal, 38,* 1, 11-26.

BFED (2002). *Annual Statistical Bulletin,* Bureau of Finance and Economic Development (BFED), Amhara National Regional State, Bahir Dar.

Bhalla, N. (2002). *Ethiopia's khat dilemma.* Retrieved August 20, 2002, from http://news.bbc.co.uk/2/hi/africa/220489/khat.htm

Bishaw, B. (1993). *Determining options for agroforestry systems for Harerghe Highlands, Ethiopia*. Ph.D. dissertation. Oregon State University.

Bojo, J. & Cassells, D. (1995). Land degradation and rehabilitation in Ethiopia. A reassessment. *Working Paper 17*. Washington D.C.: Africa Region Technical Department, Environment Division.

BOPED. (1998). Southern Nations Nationalities and Peoples' Regional Government: A Socio-Economic Profile, Bureau of Planning and Economic Development (BOPED), Awasa, Ethiopia.

Boserup, E. (1965). *The Conditions of Agricultural Growth: The Economics of Agrarian Change Under Population Pressure*. Chicago: Aldine Publishing Company.

Brooke, C. (1958). The Durra Complex in the Central Highlands of Ethiopia. *Economic Botany, 12*, 2, 192-204.

Brooke, C. (1959). The Rural Village in the Ethiopian Highlands. *Geographical Review, 49, 1,* 58-75.

Brooke, C. (1960). Khat (Catha edulis): Its production and trade in the Middle East. *Geogaphical Journal, 126,* 1, 52-59.

Brooks, M. (1997). *Trade in Africa. TED Case Studies. Qat trade in Africa.* Available at: Gurukul.ucc.american.edu/ted/QAT.HTM (accessed 2 may 2001).

Burton, R. (1987). *First Footsteps in East Africa, Or An Exploration of Harar* (Vol. 2). New York, Praeger, 1966.

CACC (2003a). *Ethiopian Agricultural Sample Enumeration, 2001/02: Results for Amhara Region, Statistical Report on Area and Production of Crops Part II.B*, Central Agricultural Census Commission (CACC), Federal Democratic Republic of Ethiopia.

CACC (2003b). *Ethiopian Agricultural Sample Enumeration, 2001/02: Results at Country Level: Statistical Report on Socio-Economic Characteristics of the Population in Agricultural Households, land, Use, and Area and Production of Crops Part I*, Central Agricultural Census Commission (AACC), Federal Democratic Republic of Ethiopia.

Carmichael, T. (2000). "Discussing the Leaf of Allah: Linguistic Aspects of Qat Culture in Harar, Ethiopia," *Ufahamu*, 28, 1, 43-69.

Carothers, J. C. (1945). Miraa as a Cause of Insanity. *East African Medical Journal*, 22, 1, 4-6.

Carrier, N. (2005). 'Miraa is Cool': the cultural importance of miraa (khat) for Tigania and Ingembe youth in Kenya. *Journal of African Cultural Studies* 17, 2, 201-218.

Carrier, N. (2005). The Need for Speed: Contrasting Timeframe in the Social Life of Kenyan Miraa. *Africa*, 75, 4, 539-558.

CBDSZ (1997). *Short Study on Khat Production and its Problems in Bahir Dar City* (in Amharic), Council of Bahir Dar Special Zone Office, Amhara National Regional State.

Clapham, C. (1988). *Transformation and Continuity in Revolutionary Ethiopia*. Cambridge: Cambridge University Press.

Conway, G. R. (1991). Sustainability in Agricultural Development: Trade-offs with Productivity, Stability and Equitability. *Paper presented at 11th Annual Symposium of the International Association of Farming Systems Research and Extension*. East Lansing, Michigan: Michigan State University.

CSA, Central Statistical Authority (1975). *Results of the National Sample Survey, Second Round: Land Area and Utilization 5*. Addis Ababa: Central Statistical Authority.

CSA, Central Statistical Authority (2004a). *Agricultural Sample Survey 2003/2004. Area and Production of Crops*, Addis Ababa.

CSA, Central Statistical Authority (2004b). *Agricultural Sample Survey 2003-04. Land Utilization*. Addis Ababa.

CSA, Central Statistical Authority (2005). *Agricultural Sample Survey 2004-05*. Addis Ababa.

CSA, Central Statistical Authority (2008). *Agricultural Sample Survey 2007-08*. Addis Ababa.

CSA, Central Statistical Authority. (2003). *Ethiopian Agricualtural Sample Enumeration, 2001/02: Results for the Oromia Region.* Statistical Report on Crop Utilization, Part III C., Addis Ababa.

CSA, Central Statistical Authroity (1999). *Agicultural Sample Survey 1998-99.* Addis Ababa.

de Haan, C. S. (1997). *Livestock and the Environment: Finding a Balance.* Brussels, Belgium: European Commission Directorate-General for Development.

Department of International Development and Cooperation, DIDC. (1999). *Sustainable livelihoods guidance sheets.* Retrieved March 1999, from http://www.livelihoods.org

Dessie, G. & Kinlund, P. (2008). Khat expansion and forest decline in Wondo Genet, Ethiopia. *Geografiska Annaler: Series B, Human Geography, 90,* 2, 187-203.

Dhaifalah, I. Š. (2004). Khat habit and its health effect. A natural amphetamine. *Biomedical Papers, 148,* 1, 11-15.

Dizikes, C. (2009, January 03). Khat -- is it more coffee or cocaine?: The narcotic leaf is a time-honored tradition in Africa but illegal in the U.S., where demand is growing. *Los Angeles Times* .

Donham, D. (1999). *Marxist Modern: An Ethnographic History of the Ethiopian Revolution.* Berkeley: University of California Press.

Eastern Ethiopia Planning Office. (1988). *Bulletin of the Socioeconomic Conditions in the Zone.* 8, Appendix 9 & 10. Harer.

Editorial, (2000, January 24). Khat has become the backbone of our country's economy. *Moresh,* 93.

Elamin, E. M. (2003). Commercialization and Food Security, Can They Go Together for the Sudanese Agrarian Economy? *10th Annual Conference of the Economic Research Forum,* 16–18 December, 2003. Marrakech, Morocco.

Elmi, A.S, Ahmed, Y.H. & Samatar, M.S. (1987) *Experience in the control of khat chewing in Somalia.* Available at: www.odccp.org/odccp/bulletin/bulletin_1987-01-01_2_page006.html (accessed 27 Feb 2003).

El-Shoura, S. M., Abdel Aziz, M., Ali, M.E., El-Said, M.M., Ali, K. Z. M. & Kemeir, M.A. (1995). Deleterious effects of khat addiction on semen parameters and sperm ultrastructure. *HumanReproduction, 10*, 9, 2295-2300

Ethiopian Customs and Excise Authority (ECEA). (1960-1991). *Annual External Trade Statistics.* Addis Ababa.

Export Facilitation and Support Committee. (2001). Overview of export and contraband trade in eastern Ethiopia: A study of challenges and solutions. Export Facilitation and Support Committee (EFSC). *Paper presented at the Workshop on Export Promotion of the Five Adjacent Regional States.* Dire Dawa.

FAO, Food and Agricultural Organization. (2004). On-line Statistical Database, Food and Agricultural Organization.

FAO, Food and Agricultural Organization. (1984). *Ethiopian Highlands Reclamation Study (EHRS). Vols. 1–2.* Rome: Food and Agricultural Organization.

FAO, Food and Agricultural Organization. (1995, June). Agricultural Production and Diversification Programmes: Food and Cash Crops. *Country Information Brief.* Rome: Food and Agricultural Organization.

FDRE, Federal Democratic Republic of Ethiopia. (1996). Food Security Strategy. *Paper Prepared for the Consultative Group Meeting of December 10-12.* Addis Ababa.

Feyisa, T.L. & Aune, J.B. (2003). Khat expansion in the Ethiopian highlands: effects on the farming systems in Habro District. *Mountain Research and Development* 23, 2, 185-189.

Ficquet, É. (2003). "Entrelacs de l'islam et du christianisme en Éthiopie" in *Cooperazione, sviluppo e rapporti con l'islam nel Corno d'Africa.* Roma, Istituto Italiano per l'Africa e l'Oriente.

Ficquet, É. (2006). "The Enjoyment and Trade of Coffee: History of Reciprocal Exchanges between Ethiopia and the Islamic Civilization." In Abdu B.K. Kasozi and Sadik Ünay (eds.), *Proceedings of the International Symposium on Islamic Civilisation in Eastern*

Africa. Kampala, Uganda, 15-17 December 2003. Istanbul: Organisation of the Islamic Conference [OIC] and Research Centre for Islamic History, Art and Culture [IRCICA].

Fitzgerald, J. (2009, March). Khat: a literature review. *Report for the Centre for Culture, Ethnicity and Health*. Retrieved 16 June 2009 from http://www.ceh.org.au/downloads/Khat_report_FINAL.pdf

Gardiner, S. (2006, November 22). That Darned Khat: In Search of New York City's Most Illusive Drug. *The Village Voice*. Retrived 2 February 2009 from http://www.villagevoice.com/contents/printVersion/206400

Gashaw, M. (1996). *The Cultivation and Use of Chat Among the Oromo of Harar with Particular Reference to Haromaya Woreda*. M.A. thesis. Department of Anthropology, Addis Ababa University.

Gebissa, E. (1994). A Preliminary Report on the Production and Exchange of Chat in the Hararghe Highlands. *IES Bullentin*, 3, 15-20.

Gebissa, E. (1997). "Consuming Leaves: the Commodification of Khat in Hararge, ca. 1930-1964." In K. Fukui, E. Kurimoto and M. Shigeta, eds., *Ethiopia in Broader Perspective: Papers of the XIIIth International Conference of Ethiopian Studies* (Vol. I, pp. 111-127). Kyoto: Shokado Book Sellers.

Gebissa, E. (1997). *Consumption, Contraband and Commodification: A History of Khat in Harerghe, Ethiopia, C. 1930-1991*. Ph.D. diss. Michigan State University. East Lansing, Michigan.

Gebissa, E. (2003). "Čat." In Siegbert Uhlig (ed.), *Encyclopaedia Aethiopica* (Vol. I). Wiesbaden, Harrassowitz Verlag.

Gebissa, E. (2004). *Leaf of Allah: Khat & Agricultural Transformation in Hararge, Ethiopia 1875-1991*. Oxford: James Currey; Athens: Ohio University Press.

Gebissa, E. (2008). Scourge of life or an economic lifeline: Public discourses on khat (*catha edulis*) in Ethiopia. *Substance Use and Misuse*, 43, 6, 784-802.

Gebresellassie, S. (2006). Intensification of Smallholder Agriculture: Option and Scenarios. *Paper prepared for the Future Agricultures Consortium* 20-22 March. The Hague: Institute of Development Studies.

Gelaw, Y. & Haile-Amlak, A. (2004). Khat chewing and its socio-demographic correlates among the staff of Jimma University. *Ethiopian Journal of Health Development, 18,* 3, 179-184.

Getahun, A. & Krikorian, A. (1973) Khat: Coffee's Rival from Harar, Ethiopia, I. Botany, Cultivation and Use. *Economic Botany, 27,* 4, 353-377.

Getahun, A. (1980). Agro-climate and agricultural systems in Ethiopia. *Agricultural Systems, 5,* 1, 39-50.

Giannini, A. J., Burge H., Shaheen, J.M., Price, W.A. (1986). Khat: Another drug abuse? *Journal of Psychoactive Drugs, 18,* 2, 155-8.

Gizaw, N. (1987). Policy options towards rapid development of the Ethiopian livestock industry. *IAR Proceedings. First National Livestock Improvement Conference, 11-13 February.* Addis Ababa.

Gondola, D. (1999). Dream and Drama: The search for elegance among Congolese youth. *African Studies Review* 42, 1, 23-43.

Govereh, J. & S. Jayne (2003). Cash cropping and food crop productivity: synergies or trade-offs? *Agricultural Economics, 28,* 1, 39-50.

Green, R. H. (1999). Khatt & the realities of Somalis: Historic, social, household, political & economic. *Review of African Political Economy, 79,* 33-49.

Griffiths, P. (1998). *Qat use in London: a study of khat use among a sample of Somalis living in London,* Central Drug Prevention Unit, Home Office, London: UK.

Gudeta, Z. & Kahssay, M. (1998, August). Competition for Scarce Land between Food and Cash Crops: The Case of Alemaya Woreda, Eastern Oromia, Ethiopia. Unpublished Paper. Department of Agricultural Economics, Alemaya University of Agriculture.

Hagos, T. (1963). Chat in the Ethiopian Economy. Unpublished mss.

Hailu, D. (2005). "Supporting a Nation: Khat Farming and Livelihoods in Ethiopia", *Drug and Alcohol Today*, 5, 3, 22-24.

Hailu, D. (2007). Should Khat be Banned? The Development Impact, *One Pager No. 40*, International Poverty Centre (IPC), now the International Policy centre for Inclusive Growth (IPC-IPG), United Nations Development Programme (UNDP)

Halbach, H. (1979). Khat—The problem today. In L. Harris (ed.), *Problems of Drug Dependence* (Research Monograph No. 27) (pp. 318-319). Washington, DC: U.S. Government Printing Office.

Hamilton, S. & Fisher, E. (2003). Non-Traditional Agricultural Exports in Highland Guatemala: Understandings of Risk and Perceptions of Change. *Latin American Research Review*, 38, 3, 82-110.

Harris, C. W. (1844). *The Highlands of Aethiopia* (Vol. 3). London: Longman, Brown & Green.

Harris, L. S., (ed.) (1966). *Problems of Drug Dependence*. Washington, DC: U.S. Government Printing Office.

Hassan, N. A., Gunaid, A.A., Abdo-Rabbo, A.A., Abdel-Kader, Z. Y., al-Mansoob, M.A. & Awad, A.Y. (2000). The effect of qat chewing on blood pressure & heart rate in healthy volunteers. *Tropical Doctor*, 30, 2, 107-8.

Hawando, T. (1982). *Summary results of Soil Science Research Program, Harerghe Highlands, Eastern Ethiopia*. Mimeo. Alemaya College of Agriculture, Alemaya, Ethiopia.

Heisch, R. B. (1945). A Case of Poisoning by Catha Edulis. *East African Medical Journal*, 22, 1, 7-9.

Hoben, Allan (1970). Social Stratification in Traditional Amhara Society. In Tuden & L. Plotnicov, eds. *Social Stratification in Africa* (pp. 187-224). New York: The Free Press.

Holland, J. H. (1995). *Hidden Order: How Adaptation Builds Complexity*. Reading: Addison Wesley, UK. http://www.unu.edu/unupress/food/8F083e/8F083E07.htm

Huntingford, G. W. (1965). *The glorious victories of Amda Syon, King of Ethiopia*. Oxford: Clarendon Press.

Isenberg, C. W. & J. L. Krapf, (1843). *The Journals of the Rev. Messrs Isenberg and Krapf* London, Seeley Burnside and Seeley.

Jaenen, C. (1956). The Galla or Oromo of East Africa. *Southwestern Journal of Anthropology, 12,* 2, 171-190.

Kalix, P. (1987). Khat: Scientific knowledge and policy issues. *British Journal of Addiction, 82,* 1, 47-53.

Kassa, H. (1986). *Livestock production, household food security and sustainability in smallholder mixied farms. A case study from Kombolcha Woreda, of Eastern Ethiopia*. Dire Dawa, Ethiopia: Department of Agricultural Economics. Alemaya University.

Kassa, H. (2000). *Livestock production, household food security and sustainability in smallholder mixed farms: A case study from Komolcha Woreda of Eastern Ethiopia*. M. A. Thesis. Uppsala: Swedish University of Agricultural Sciences. Department of Rural Development.

Kassa, H., Blake, R. W. & Nicholson, C. F. (2002). The crop-livestock subsystem and livelihood dynamics in the Harar Highlands of Ethiopia. *International Conference "Responding to the Increasing Global Demand for Animal Products."* Marida, Mexico: British Society of Animal Science.

Kassaye, M., Sherief, H.T., Fissehaye, G., & Teklu, T. (1999). Drug use among high school students in Addis Ababa and Butajira. *Ethiopian Journal of Health Development, 13,* 2, 101-6.

Kebede, D., Alem. A., Mitike. G., Enquoselassie, F., Berhane, F. & Abebe, Y. (2005). *Khat and alcohol use and risky sex behaviour among in-school and out-of-school youth in Ethiopia.* Retrieved August 22, 2006, from http://www.biomedcentral.com/1471-2458/5/109

Kebede, Y. (2002). Cigarette smoking and khat chewing among college students in north west Ethiopia. *Ethiopian Journal of Health Development, 16,* 1, 9-17.

Kedir, A. (c. 2005). *Socioeconomic impact of export oriented agricultural production on farmers, in eastern Ethiopia.* Retrieved April 15, 2006, from http://www.isser.org/3%20adem.pdf

Kennedy, J. (1987). *The Flower of Paradise: Institutionalised use of the drug khat in North Yemen.* Dordrecht, The Netherlands: D. Reidel Press.

Kennedy, J. Teague, J., Rokaw, W. & Conney, E. (1983). A medical evaluation of the use of khat in North Yemen. *Social Science & Medicine, 17,* 12, 783-93.

Klein, A. & Beckerleg, S. (2007). Building Castles of Spit: The Role of Khat in Work, Ritual, and Leisure. In J. Goodman, P. Lovejoy, & A. Sherratt, eds. *Consuming Habits: Global and Historical Perspectives on How Cultures Define Drugs* (pp. 238-354). London: Routledge.

Klingele, R. (1998). *Hararghe Farmers on the Crossroad between Subsistence and Cash Economy.* Unpublished report, UNDP-Emergency Unit for Ethiopia, Addis Ababa, Ethiopia. United Nations Emergency Unit for Ethiopia.

Krikorian, A. D. (1984). Kat and Its Use: An Historical Perspective. *Journal of Ethnopharmacology, 12,* 115-178.

Kydd, J. (2002). Agriculture and Rural Livelihoods: Is Globalization Opening or Blocking Paths out of Rural Poverty? *AgREN Paper No. 121. Agricultural Research and Extension Network.* ODI. London.

Langlais, C., Weil, M., & Wibaux, H. (1984). *Farming Systems Research Preliminary Survey and Future Program.* Department of Agricultural Economics, French Technical Cooperation, Dire Dawa, Ethiopia.

Lemessa, D. (2002). *Migrants cause potential social and environmental crisis in Bale.* A Joint Mission by the UN-Emergencies Unit for Ethiopia with the Ethiopian Evangelical Church Mekane Yesus and the Oromiya Regional Government Field Assessment

Mission, United Nations Emergencies Unit for Ethiopia, Addis Ababa.

Letters from an anti-khat group. (2004, October). (personal communication with the author.

Luqman, W. D. (1976). The use of khat (Catha edulis) in Yemen: Social and Medical Observations. *Annals of Internal Medicine, 85*, 2, 246-249.

Mains, D. (2004). Drinking, Rumor and Ethnicity in Jimma, Ethiopia. *Africa*, 73, 4, 341-360.

Mains, D. (2007). "*We are just sitting and waiting*": Aspirations, Status, and Unemployment among Young Men in Jimma, Ethiopia. Ph.D. Dissertation. Emory University: Department of Anthropology.

Marrassini, P. (Ed. & Trans.) (1993). *Lo Scettro e la Croce: La Campgna di 'Amda Seyon contro l'Ifat (1332)*. Napoli: Istituto Universitario Orientale.

Merab, P. (1912). *Medecins et Medecine en Ethiopie. Generalites Pathologie Medicale, Pathologie, Chirurgicale et Accouchements, Medecins Etrangers en Ethiopie*. Paris: Vigot Freres Edit.

Mercier, J. (c. 1980-1982). "un mythe éthiopien d'origine du café et du khat," *Abbay*, 11.

Moktar, M. (1876). Notes sur le pays de Harrer. *Bulletin de la société Khédiviale de Géographie,* 1, 5, 369-372.

Mulatu, E. & Kassa, H. (2001). Evolution of Smallholder Mixed Farming Systems in the Harar Highlands of Ethiopia: The Shift Towards Trees and Shrubs. *Journal of Sustainable Agriculture*, 18, 4, 81-112.

Mulatu, E., Ibrahim, O. E. & Bekele, E. (2005). Policy Changes to Improve Vegetable Production and Seed Supply In Hararghe, Eastern Ethiopia. *Journal of Vegetable Science*, 11, 2, 81-206.

National Bank of Ethiopia, NBE. (2004). *Annual Report 2003/04*. Ethiopian Fiscal Year 1996, National Bank of Ethiopia, Addis Ababa.

National Bank of Ethiopia, NBE. (2005). *Quarterly Bulletin*, Addis Ababa, Ethiopia.

Negatu, D. (2004, August). Chewing Over Khat Closures. *Fortune*, August 1: 1-3, 21.

Nencini, P. & Ahmed, A.M. (1989). Khat consumption: A pharmacological review. *Drug Alcohol Dependence*, 23, 1, 19-29, .

Newell, S. (2005). Migratory Modernity and the Cosmology of Consumption in Cote d'Ivoire . In L. Trager, (ed.) Migration and Economy: Global and Local Dynamics (pp. 163-192). Altamira Press: Walnut Creek.

Nicholson, C.F., Blake, R.W. & Lee, D.R. (1995). Livestock, deforestation and policy making: Intensifying cattle production systems in Central America revisited. *Journal of Dairy Science*, 78, 3, 719-734.

Nicholson, C.F., Blake, R.W., Reid, R.S. & Schelhas, J. (2001). Environmental impacts of livestock in the developing world. *Environment*, 43, 2, 7-17.

Nida, W. (2004). Cultures of Chat Consumption in Addis Ababa and its Environs: Discussing Some Contesting Discoures on the Meaning and Implications of Such Cultural Practices. Paper presented to the *National Workshop on Khat: Consumption, Production, and Trade*. Addis Ababa: Institute of Ethiopian Studies.

Odenwald, M. (2007). Chronic khat use and psychotic disorders: A review of the literature and future prospects. *SUCHT*, 53, 1, 9-22.

Pankhurst, R. (1997). The Coffee Ceremony and the History of Coffee Consumption in Ethiopia. In K. Fukui, E. Kurimoto & M. Shigeta, eds., *Ethiopia in Broader Perspective: Papers of the XIIIth International Conference of Ethiopian Studies*, (Vol. II. pp. 516-539). Kyoto: Shokado Book Sellers.

Pankhurst, R. K. (2002). Across the Red Sea: Ethiopia's Historic Ties with Yaman, *Africa*, 47, 3, 393-419.

Parsons, T. (1951). *The Social System*. Glencoe: Free Press.

PEDD (1998) Gurage Zone Socio-Economic Base-Line Survey, Planning and Economic Development Department, Wolkite.

Pennings, E.J.M., Opperhuizen, A., & van Amesterdam, J.G.C. (2008). Risk assessment of khat use in the Netherlands: A review based on adverse health effects, prevalence, criminal involvement and public order. *Regulatory Toxicology and Pharmacology, 52*, 3, 199-207.

Pentelis, C. H. (1989). Use and abuse of khat (Catha edulis): A review of the distribution, pharmacology, side effects and a description of psychosis attributed to khat-chewing. *Psychological Medicine, 19*, 3, 657-668.

Piguet, F. (2002). *Assessment Field Trip to East and West Hararghe Zones (Oromiya Region)*. Field Assessment Report, 3-9 September 2002. Addis Ababa: United Nations Development Program-Emergencies Unit for Ethiopia.

Piguet, F. (2003). *Hararghe Food Security Hampered by Long Term Drought Conditions and Economic Constraints*. United Nations Emergiencies Unit for Ethiopia, Addis Ababa.

Plan for Accelerated and Sustainable Development in Ethiopia (PASDEP). (2007). Plan for Accelerated and Sustainable Development in Ethiopia. Ministry of Finance and Economic Development, Addis Ababa, Ethiopia.

Poluha, E. (2004). *The Power of Continuity: Ethiopia through the Eyes of its Children*. Uppsala: Nordiska Afrikainstitutet.

Poschen, P. (1986). An evaluation of the *Acacia albida*-based Agroforestry practices in the Hararghe highlands of Eastern Ethiopia. *Agroforestry Systems, 4*, 129-143.

Poschen, P. (1987). The Application of Farming Systems Research to Community Forestry – A Case Study in the Harerghe Highlands, Eastern Ethiopia. In D. Knuth (ed.), *Tropical Agriculture* (Vol. 1, pp, 1-250). TRIOPS. Verlag: Langen, Germany.

Quiñones, M.A., Borlaug, N.E. & Dowsell C.R. (1997). A Fertilizer-based Green Revolution for Africa. In Buresh, R.J., Sanchez, P.A. &

Calhoun, F., (eds). *Replenishing Soil Fertility in Africa* (pp. 81-95). Madison: Wisconsin, USA.

Rafn, I. K. (2002). *Cash crops in combination with food crops stimulate sustainable agriculture and improved livelihoods in Africa.* Retrieved April 12, 2006, from http://www.sustdev.org

Raja'a, Y. A., Norman, T.A., al-Warafi, A.K., al-Mashraki, N.A. & al-Yosofi, A.M. (2001). Khat chewing is a risk factor of duodenal ulcer. *East Mediterranean Health Journal*, 7, 3, 568-70.

Rawlins, M. (2005). *Khat (Qat): Assessment of Risk to the Individual and Communities in the UK.* Advisory Council on the Misuse of Drugs. The Home Office, UK.

Reijntjes, C., Haverkort, B. & Waters-Bayer, A. (1992). *Farming for the Future: An Introduction to Low-external Input and Sustainable Agriculture.* Information Centre for Low-External-Input and Sustainable Agriculture (ILEIA). London: The Macmillan Press Ltd.,.

Risoud, G. (1987). *Evolution of Peasant Agriculture in the Eastern Harerge Highlands: Development Prospects.* Farming Systems Research. Alemaya: Alemaya University of Agriculture.

Ruthenberg, H. (1980). *Farming Systems in the Tropics.* Third Edition. Oxford: Oxford University Press.

Saeed, K. (1994). *Development Planning and Policy Design: A System Dynamics Approach.* Hants (England) and Vermont (USA): Ashgate Publishing.

Save the Children Fund (UK)–Ethiopia. (1996). *Food Economy Profiles, Risk Mapping Project. Hararghe, Ethiopia.* Unpublished Report.

Serneels, P. (2007). The Nature of Unemployment among Young Men in Urban Ethiopia. *Review of Development Economics,* 11,1, 170-186.

Seyoum, E. Kidane, Y. & Gebru, H. (1986). Preliminary study of income and nutritional status indicators in two Ethiopian Communities. *Food and Nutrition Bulletin,* 8, 37, 37-41.

Shash, Z. (2001). Shisha and Shisha Houses, *Salafiya*. A monthly Islamic magazine that is still published in Addis Abäba, vol.1, October/November. [in Amharic].

Sherif, M. et al. (2002). Towards the Formulation of a Comprehensive Q'at Policy in the Republic of Yemen: Technical Field Study. *National Conference on Q'at*, 6-7 April.

Simoons, F. J. (1960). *Northwest Ethiopia: Peoples and Economy*. Madison: University of Wisconsin Press.

Stage, O. & Rekve, P. (1998). Food Security and Food Self-sufficiency: The Economic Strategies of Peasants in Eastern Ethiopia. *European Journal of Development Research, 10*,1, 189-200.

Start, D, Tefera, T. L. Kjell, P. & Tesfaye, M. (2005). *Choices and constraints: a study of livelihoods in the eastern Ethiopian highlands*. Addis Ababa: Ethiopia

Stenhouse, P. L. (Trans.). (2003). *Futūh al-Habaša, The Conquest of Abyssinia*. Hollywood, CA: Tsehai Publishers and Distributors.

Sterman, J. D. (2000). *Business Dynamics: Systems Thinking and Modeling for a Complex World*. Boston: Irwin/McGraw Hill, USA.

Storck, H., Emana, B., Adenew, B., Borowiecki A. & W/Hawariat, S. (1991). *Farming Systems and Farm Management Practices of Smallholders in Hararghe Highlands. Farming Systems and Resource Eonomics in the Tropics* (Vol. 11). Kiel: Wissenschafts Verlag.

Szendrei, K. (1980). The chemistry of khat. *Bulletin on Narcotics, Special Issue Devoted to Catha edulis (khat), 32,* 3, 5-35.

Tadesse, E. (1958). Preparation of Tag among the Amhara of Sawa. *Ethnological Society Bulletin, 8,* 101-109.

Tefera, T.L. (2003). *Livelihood strategies in the context of population pressure: a case study in the Hararghe highlands, eastern Ethiopia*, PhD thesis, Department of Agricultural Economics, Extension and Rural Development, University of Pretoria.

Tefera, T.L., Kirsten, J.F. & Perret, S. (2003). Market incentives, farmers' response and a policy dilemma: the case study of expansion of chat production in the Hararghe highlands, eastern Ethiopia. *Agrekon*, 42, 3, 213-227.

Tefera, T.L., Perret, S. and Kirsten, J.F. (2005). Diversity in livelihoods and farmers' strategies in the Hararghe highlands, eastern Ethiopia, *International Journal of Agricultural Sustainability*, 2, 2, 133-146.

Tesfaye, A. (1957). The Funeral Custom of the Kottu of Hararge. *University College of Addis Ababa Ethnological Society Bulletin*, 7, 35-40.

United Nations Office on Drugs and Crime, UNODC. (1996). *Amphetamine-type stimulants: A global view*. Retrieved March 26, 2008, from http://www.unodc.org/undoc/technical_series_1996-0101_1.html

Varisco, Daniel (1986). On the Meaning of Chewing: The Significance of qat in the Yemen-Arab-Republic. *International Journal of Middle East Studies*, 18, 1, 1-13.

Vosti, S. A. (1995). Sustainability, Growth, and Poverty Alleviation: Research and Policy Issues for Tropical Moist Forests. In A. S. Juo & Freed, R.D. (eds.), *Agriculture and Environment: Bridging Food Production and Environmental Protection in Developing Countries. Soil Science Society of America*. Madison: American Society of Agronomy.

Wakjira, M. (1989). *A study of erosion and conservation on the Alemaya catchment*. M.Sc. thesis. Department of Agricultural Economics, Alemaya University of Agriculture.

Wakjira, M. (2004). The Impact of Khat on Household, Regional and National Economy: A Study of Eastern Ethiopia. In E. Gebissa ed., *Proceedings of the National Workshop on Khat* (pp. 1-21). Addis Ababa: Institute of Ethiopian Studies.

Walta Information Center (2008). Nation Exports 22,390 tons of Khat. http://www.waltainfo.com/. Accessed 9 July 2009.

Weir, S. (1985). *Qat in Yemen: Consumption and Social Change*. London: British Museum Press.

Bibliography

Westphal, E. 1975. *Agricultural Systems in Ethiopia*. Agricultural Research Reports 826. Centre for Agricultural Publishing and Documentation. Wageningen, The Netherlands: Centre for Agricultural Publishing and Documentation.

Wibaux, H. (1986). *Agriuclture in the Highlands of Hararghe, Kombolcha Area: Study of Six Farms*. Dire Dawa: Farming Systems Research, Alemaya Univerisity of Agriculture.

Woldemichael, W. (2003, December 26). Health hazards associated with khat consumption. *Addis Tribune*, pp. 8-9.

Woldemichael, W. (2004, 7 May). The chat problem: A reaction to Henoke Semaegzer's article in the Reporter. *Addis Reporter*.

World Bank. (2004). *Human Development Index*.

World Commission on Environment and Development, WCED. (1987). *Our Common Future*. Oxford: Oxford University Press.

World Helath Organization, WHO. (2003). Expert Committee on Drug Dependence, Thirty-Third Report, World Health Organization Technical Report Service 915.

Yousef, G., Huq, Z. & Lambert, T. (1995). Khat chewing as a cause of psychosis. *British Journal of Hospital Medicine*, 54, 7, 322-6.

Zeleke, W. A. (2004, November 19). Crux of the matter: The emerging chat (khat) business. *Addis Tribune*, pp. 4, 9.

Zinberg, N. (1984). *Drug, Set and Setting: The Basis of Controlled Intoxicant Use*. New Haven: Yale University Press.

Zola, I. K. (1972). Medicine as an instituion of social control. *Sociological Review*, 20, 4, 487-504.

Notes on Contributors

&

Hussein Ahmed† (PhD, History, Manchester) was Professor in the Department of History at Addis Ababa University. He is the author of *Islam in Nineteenth-Century Wallo, Ethiopia: Revival, Reform and Reaction* (Leiden, 2001) and several articles on Islam in mediaeval and contemporary Ethiopia. Recent publications include "Waqf-Land in Nineteenth-Century Wallo (Ethiopia)" in Donald Crummey (ed.), *Land, Literacy and the State in Sudanic Africa* (2005) and "Coexistence and/or Confrontation? Towards a Reappraisal of Christian-Muslim Encounter in Contemporary Ethiopia," *Journal of Religion in Africa*, 36, 1 (2006).

Christopher Clapham (PhD Politics) is at the African Studies Center at Cambridge University. He has widely published on African politics, including *Haile-Sellassie Government* (1969) and *Transformation and Continuity in Revolutionary Ethiopia* (1998).

Ezekiel Gebissa, (PhD History, Michigan State), is Associate Professor of History at Kettering University. He is the author of *Leaf of Allah: Khat and the Transformation of Agriculture in Harerge Ethiopia, 1875–1991* (James Currey and Ohio University Press, 2004) and editor of *Contested Terrain: Essays on Oromo Studies, Ethiopianist Discourses, and Politically Engaged Scholarship* (Red Sea Press, 2008). He has published several articles in refereed journals. He is the current editor of the *Journal of Oromo Studies*.

Degol Hailu (PhD Economics, SOAS) is Acting Director of the UNDP's International Policy Centre for Inclusive Growth (formerly known as the *International Poverty Centre*) in Brazil. He is the co-author of *The Khat Controversy: Stimulating the Debate on Drugs* (Berg, 2007).

Habtemariam Kassa (PhD Livelihoods and Rural Development, Swedish University of Agricultural Sciences) is regional scientist in the Forests and Livelihoods Program of the Center for International Forestry Research (CIFR) in Addis Ababa. He has published articles in peer reviewed journals on the evolution of farming systems in the Harer Highlands of Ethiopia. His current work focuses on livelihood strategies and on technical and institutional innovations for management of natural resources in arid and semi-arid areas of Africa. Before joining CIFR, he had taught for many years at Haramaya University, Ethiopia.

Daniel Mains (PhD Anthropology, Emory) is a Faculty Fellow in the Department of Anthropology at Colby College. He has published articles in the journals *Africa* and *American Ethnologist*. He is currently working on a collaborative research project funded by the National Science Foundation, titled "Poverty, Social Change, and Shifting Expectations: The Makings of Mental Health Disorders among Ethiopian Adolescents."

Tesfaye Lemma Tefera (PhD Rural Development, Pretoria) is Assistant Professor in Department of Rural Development and Agricultural Extension at Haramaya University. He has published articles in peer reviewed journals on tural livelihoods, khat production and food and nutritional security.

David Start is with the New Economic Foundation, UK.

Index

abba gar 17-19
Abeshege (Goro) 134
abstinence syndrome 83
Adama 133
Adamitulu 61
Addis Ababa xi, 8, 25-27, 51, 60, 62, 85, 112, 133, 134, 136, 144, 158, 159, 168, 176
Aden 26, 90, 93, 96, 98, 99, 193
Advisory Council on Misuses of Drugs 168
Afar Regional State 133
Afghanistan 82
Africa 1, 2, 14, 48, 82, 91, 96, 119, 167, 193, 201, 205
Agaro 34, 61
Agricultural Development Led Industrialization (ADLI) 124, 128, 129, 131, 146, 202, 203
agriculture (general) 3, 6, 10, 57, 91-96, 102, 104, 107, 108, 117, 119, 124, 128, 129, 131, 132, 146, 151, 152, 165, 181, 185, 193, 207
  cash cropping 3, 4, 90, 96, 104, 106, 113, 114, 118, 119, 125, 180, 183, 197
  diversification 96, 100, 107, 108, 118, 124, 151, 155, 158, 163, 164, 182, 186
  drought 93, 138, 139, 151, 165, 168, 170, 172, 175, 178, 179, 184, 185, 206
  extensification 3
  intensification 9, 104, 124, 150, 155, 166, 180, 181, 184, 197
  intercropping 3, 96, 100, 105, 178, 185, 192
agro-ecological zones (AEZ) 171, 173

Ahmad b. Ibrāhīm or Grañ, *Imām* 14
Amaro 137
Ambassal 14
Amde Seyon 57
Amhara Regional State 8, 138, 139
Amhara State University 144
amphetamine 38, 81, 83, 168
Anti-khat group 81
Aqara 22, 109, 112
Arabian Peninsula 1
Arafa 64
Asfa Wossen Haile Sellassie 16
Atlanta 46, 47
Australia 2, 78
Awassa 137, 158
Aweday 110-112, 120, 122, 159, 162
Awracha 17

Badessa 109-111
Bahir Dar 8, 62, 127, 141, 142, 144, 145, 158
Bahir Dar University 142
barrak bay 17
barsa sadi 19
bartcha 18, 59, 69
basmala 17
Beke, Charles 58
Bench-Maji 132, 137

Benishangul-Gumuz 133
British Somaliland 193
Bure 141
Burton, Richard 1, 2, 58

Campaign Through Cooperation 76
Canada 2, 10, 167, 192
Cathinone/cathine 81-83, 108, 167, 202, 205
Cereals 93, 130, 140, 141, 146, 157, 202
chabsii 38, 39
Cheha 134
Chercher 63, 92, 99, 106, 118
Chilalo Agricultural Development Unit (CADU) 202
Christians 1, 5, 23, 34, 35, 57
cocaine 70, 71, 82, 203, 204, 206
cocoa 206
Coffee 3, 4, 17-19, 21, 23, 36, 37, 40, 42, 45, 63, 70, 76, 82, 84, 89, 90, 93, 97, 100, 102, 103, 106, 117, 118, 124, 125, 129, 131, 133, 138, 140, 141, 146, 147, 157, 160, 169, 171-173, 175, 176, 178, 179, 181, 184, 185, 192, 203, 204
Coffee Berry Disease 175, 176

Index

Colombia 141
controlled substance 81, 82, 185, 193
criminalization, khat use 147, 185
crop-livestock integration 8, 149, 151, 154, 155
Cultural drug dependence 83

Dagan 22
Dangla 141
Dase 22
Deder 9, 169-173, 176, 178, 184, 186, 187
Delgi 141
Derashe 137
Derg 25, 32, 90, 101, 103, 104, 106, 110, 111, 117, 119-122, 202, 203
development strategy 129, 150
Diaspora 131
Dinsho Khat Export Company 160
Dire Dawa 26, 34, 64, 90, 93, 98, 105, 110, 111, 120, 122, 123, 133, 160, 162, 171, 176
Djibouti 2, 23, 90, 93, 96, 98, 110, 120, 131, 147, 157, 160, 162, 167, 168, 171, 176, 178, 193
drought 93, 138, 139, 151, 165, 168, 170, 172, 175, 178, 179, 184, 185, 206

Drug Enforcement Agency (DEA) 82
du'a adragiwoch 16
duriye 47, 53, 56

East Africa 167, 193
East Harerge Zone 168-170
Eastern Harerge 95, 105-107, 118, 123, 169
embwacho (Rumex nervosus) 22
Employment Generation Scheme (EGS) 182
Endasa 141
Endebir 134
Enemor 134
Ener 134
Enset (*Ensete ventricosum* or *E. ventricosum*) 133, 141
Environmental degradation 163, 201
Ergoyye 22
Esia 134
Ethiopia x, xi, 1-4, 6-8, 10, 13, 14, 18, 19, 21, 22, 25, 29-31, 33, 34, 43, 44, 46, 48, 51, 53-55, 57-60, 62, 63, 66, 68-73, 76, 81, 83, 84, 89-92, 99, 101, 104-106, 112, 114, 117, 120-122, 124, 127-129, 131, 133, 146, 149, 151, 158-161, 165, 167-169,

176, 184, 185, 186, 191, 193, 195-199, 202
Ethiopian agriculture 3, 10
Ethiopian Airlines 90, 120, 192
Ethiopian Herald 161
Ethiopian People's Revolutionary Democratic Front (EPRDF) 201, 202
Ethiopian revolution 110
Ethio-Somali War 120
Europe 2, 118, 131, 147, 161, 167, 178, 193
export markets 130, 146, 165, 168

Farming Systems Unit 103
Farrah 53
ferdo 109
Finoteselam 141
food crop production 3, 9, 99, 104, 107, 118, 119, 180, 184
Food security vii, 3, 4, 8, 104, 112, 114, 116, 124, 150, 156, 165, 169, 181, 184, 185, 206

Gachcha 16
Gamo Goffa 137
ganfo 20
Garba 22

Gedeo 132, 137
German Agency for Technical Cooperation (GTZ) 143
gesho 83
Gunchere 134
Gurage 26, 61, 127, 132-138, 147, 148

Habro 60, 109, 157, 168, 178
Hadiya 132, 137
Haile Sellassie, Emperor 16
Hamusit 141
Haramaya 51, 98, 99, 103-111, 113, 151, 168, 180, 181
Haramaya University 51, 103, 151
Harer 1, 2, 7, 58, 93, 99, 106, 116, 120, 122, 149, 151, 165
Harer highlands 8, 9, 92, 99, 104, 118, 149, 151-153, 155-159, 161, 163-166, 199
Harerge viii, 6, 8, 59-61, 63, 65, 73, 75, 83, 90-98, 100, 102-111, 113, 117-123, 135, 157, 159, 168-171, 174, 175, 184, 193, 195
Harerge highlands 8, 59, 65, 90-93, 96, 106, 108, 113, 117, 119, 121, 122, 171, 184
heroin 71, 82

HIV/AIDS 35, 42-45, 67, 68
Horn of Africa 2, 91, 192, 200

Ifat 14, 57
International Council on Alcohol and Addictions 80
International Fund for Agricultural Development (IFAD). 143
Iraq 82
Irish potato 93, 171-173, 175, 179, 181
irrigation 9, 104, 107, 170, 172, 174, 175, 177, 179, 181-183, 186, 197
Islam 2, 14, 15, 23, 63, 64, 204
Islamic law 14
Iyasu. *Lij* 63

jabana 16
Jafjaf 17, 22
Jijiga v, 4, 120, 122, 162
Jijiga Zone 4
Jimma 7, 8, 29, 32, 34-36, 38, 41, 43, 44, 47, 48, 61, 158

kaddam 18
Katumomia 64

kebeles or neighborhood associations 76, 171, 179
Kenya 53, 66, 79, 133, 147, 161, 167
khat chewing 2, 6, 7, 8, 13, 14, 17, 21, 30, 36, 43, 58-65, 68-70, 73, 74, 76-78, 82-84, 191, 194-196
cultural/social practice 9, 39, 53, 59, 72, 74
health and social effects 54, 64, 65, 67, 79, 185, 198
highs, see *mirqana*
Muslim habit 128
prohibition x, 9, 10, 59, 73, 75, 77-81, 83, 147, 191, 193-197, 199, 204-206
ritual 13-25, 58, 73
khat cultivation viii, 2, 4, 9, 25, 62, 90, 91, 102, 104, 118, 127-129, 131, 134, 141, 156, 191, 193, 201-204
farmers 1, 3, 4, 9, 10, 60, 69, 70, 76, 89, 90, 92, 93, 95-104, 106, 107, 109, 113-119, 121, 124, 128-131, 134-136, 139, 142, 143, 146, 147, 149-159, 162-165, 168, 169, 172, 174-176, 178-181, 183, 185, 186, 192, 194, 196, 199, 202, 203
intercropping 3, 96, 100, 105, 178, 185, 191

khat exports 3, 58, 89, 90, 94, 98, 104, 106, 109-111, 114, 115, 119-123, 129-131, 133, 138, 144-146, 151, 153, 154, 156-163, 165, 166, 168, 169, 171, 176-178, 183, 185, 192-194

earnings 64, 120, 129, 146, 147, 151, 159-161, 169, 185, 192

unit price 35-37, 89, 96, 99, 102, 106-112, 114, 115, 117, 118, 128-130, 139, 141-144, 146, 152-158, 160, 161, 163, 172, 174-178, 180, 184, 185, 192, 193, 203, 206

value 4, 5, 21, 33, 41, 95, 102, 109, 112, 115, 117, 120, 121, 130, 138, 146, 157, 160, 161, 172, 174, 178-180, 183, 199, 201

volume 4, 6, 8-10, 21, 98, 106, 108, 120, 123, 137, 155, 158, 160, 161, 191, 201, 204, 205

Khat leaves 4, 17, 22, 64

cash crop 2-4, 6, 10, 14, 89, 90, 93, 95-98, 100, 102-104, 106-108, 113-116, 118, 119, 128, 168, 171-174, 176, 178, 180, 181, 183, 185, 191, 192, 197

production 4-6, 8, 9, 69, 70, 90, 91, 96, 98, 106, 112, 115, 116, 122, 129, 131, 133-135, 138, 141, 151, 158, 159, 164, 165, 167-169, 174, 180, 181, 184, 185, 197, 203

stimulant effect 2, 50, 51, 55, 67, 69, 76, 81, 141, 203

khat varieties and market brands 14, 19, 22, 109, 111, 127, 133, 144, 150, 177

awaday 122
beleche 127, 133
Colombia 127, 141, 144
daalota 110, 178
diimaa 110, 178
hamerkot 109, 178
qarti 109
uratta 109-111
wando 152

Kinbaba 141
Kokir 134
Kombolcha 22, 99, 104, 109, 111, 152, 157, 164

Landholding and fragmentation 139, 151, 152, 163, 174, 195

Las Vegas 47

Law enforcement 193

League of Nations Advisory Committee on the Traffic of Dangerous Drugs 81

Index

livestock 3, 8, 93, 94, 108, 115, 135, 149-151, 153-155, 164, 181, 182
London 176
lullu qachcha 21

Madagascar 80
Malthusian spiral 201
Marxist 32
Mawlid 64
Meher and Aklil 134
Menelik, Emperor 63
Merawi 141
Meru 14, see also mira
Meshenti 141
Meta 9, 169-173, 178, 184
Middle East 2, 90
mirqana 19, 38-50, 67, 74
Mojo 133
Moktar, Mohammed 58, 59
Moore, Sidney L. 82
Mota 141
multinational corporations 203
Muslim 1, 7, 13, 15, 20, 23, 34, 35, 55, 57-60, 63, 64, 74, 128, 138, 144, 147

Netherlands 78, 161
non-governmental organizations (NGO) 4, 96, 117, 143, 150, 155, 172

North America 2, 118, 131, 167

Oromia Regional State 61, 168, 169
Oromo people 50, 63, 71, 75, 95
Oromo ritual 15
Orthodox Christians 35

Pagumen, also *Qwagme* 16
population pressure 3, 119, 139, 150, 151, 158, 164, 184, 206
prohibitionists 73, 77, 78, 81, 83, 194, 195
psychological dependence 83

Qallu 14
Qarsa 22
Qarsa district 95
Qattataye 22
qimaha 18
Quindi 175, 176, 178-180, 184
Qur'an 14, 21

rakabot 16
Ramadan 1, 64

Republic of Somaliland 2
Rūh Anti-Khat Club 25

Second World War 21, 96, 120
self sufficiency 3, 4, 91, 96-98, 100, 104, 107, 119, 192, 202
Shashemene 158
Shawa 14
shisha 23, 36
Sidama 61, 132, 134, 135, 137, 138
Silti 137
Somaliland Protectorate 98
Somalis 75, 147, 167
Southern Nations, Nationalities, and Peoples Region of Ethiopia (SNNPR) 8, 61, 127, 133-139
Southwestern Ethiopia 39, 60
Sustainability 9, 10, 113, 151, 152, 162, 185, 191, 193, 197
Swedish International Development Agency (SIDA) 61, 127, 132-138, 143

Tahara 17
Takla Haymanot Square 25
talla 15
Tekle Dengay 141

Teklehawariat, *Fitaurari* 63
Tigray 62
Togouchale 162

Uganda 147
unemployment 7, 21, 24, 29-31, 33, 35, 43-45, 47, 48, 51, 54
United Kingdom 71, 78
United Nations 81
United Nations Commission on Narcotic Drugs 81
United Nations Office on Driugs and Crime (UNODC) 81
United States xi, 2, 10, 45, 73, 82, 167, 193
US Circuit courts 82
US Code of Federal Regulations 82

Vietnam 73

Wadaja 7, 15-21
 ya dubarti 15
 ya lej wadaja 15
 ya tolfanna 15
 ya wand 15
Wallo 7, 13-17, 20-22, 144, 145
Warrabbacho 22

Index

West Africa 206
West Harerge Zone 168
Western Europe 2
WHO Expert Committee on Drug Dependence 82
Wendo Genet 4, 127, 133, 134, 136
World Bank 112

Yajju 14
Yemeni 7, 14, 17, 22, 66, 96, 131
Yemeni Arabs 22, 96
YMCA 76

Zeghe 141
Zeleke Beyene, Colonel 63
Zemet 141
Zenzelma 141
zurba 23